T0202479

Game Theory and Fisheries Management

Lone Grønbæk · Marko Lindroos ·
Gordon Munro · Pedro Pintassilgo

Game Theory and Fisheries Management

Theory and Applications

Springer

Lone Grønbæk
Department of Business and Economics
University of Southern Denmark
Odense, Denmark

Gordon Munro
Vancouver School of Economics
University of British Columbia
Vancouver, BC, Canada

Marko Lindroos
Department of Economics
and Management
University of Helsinki
Helsinki, Finland

Pedro Pintassilgo
Faculty of Economics and CEFAGE
University of Algarve
Faro, Portugal

ISBN 978-3-030-40114-6   ISBN 978-3-030-40112-2   (eBook)
https://doi.org/10.1007/978-3-030-40112-2

This Springer imprint is published by the registered company Springer Nature Switzerland AG
The registered company address is: Gewerbestrasse 11, 6330 Cham, Switzerland

Preface

This book is the outcome of a truly cooperative repeated game among the four authors, which evolved over a long time horizon. The origins of the book can be traced back to 2007, when three of the authors came together to mount a doctoral course on game theory and fisheries management and realized that no published relevant teaching material existed. Three years later, the same three authors presented the doctoral course once more and prepared the first draft of a table of contents for the book. The idea then lay dormant for several years until 2016, when it was revived, and the Grand Coalition was expanded from three to four. It was realized that the theoretical contribution, in of and by itself, would be insufficient; that it would be essential to reach out, not only to students and academics, but to stakeholders and policymakers as well. This is the book that you now have before you. The reader is presented with a solid theoretical background, accompanied by case studies demonstrating the relevance of the theory to real-world policy issues. The policy implications of the theory plus case studies are made explicit, chapter by chapter. The chapters can be read independently, and subsections of the chapters may be of more relevance and importance to different readers.

We hope you will find the book useful! Enjoy reading!

Odense, Denmark
Helsinki, Finland
Vancouver, Canada
Faro, Portugal

Lone Grønbæk
Marko Lindroos
Gordon Munro
Pedro Pintassilgo

Acknowledgements Pedro Pintassilgo acknowledges financial support from FCT—Portuguese Foundation for Science and Technology, through project «UID/ECO/04007/2019».

Contents

Chapter 1
Introduction to the Application of Game Theory in Fisheries Management

Abstract The role of game theory (the theory of strategic interaction) in the economics of the management of world capture fisheries resources has evolved gradually over time. If the commencement of modern fisheries economics can be seen to be marked by the publication of H. Scott Gordon's seminal 1954 article, it can be said that, for the first quarter-century of modern fisheries economics, game theory played a negligible role. There was no perceptible strategic interaction, or at least none seen worth bothering about. All of this changed with the advent of the 1982 UN Convention on the Law of the Sea and the EEZ regime. Coastal states establishing EEZs were forced to recognize that some of the fisheries encompassed thereby would be shared with other states. Since strategic interaction between and among states sharing these international fishery resources is central to the management of the resources, fishery economists analysing such management had no choice but to bring game theory to bear. The steadily increasing complexity of international fishery management brought forth the need for game theory models of greater and greater sophistication. The application of game theory to the economics of the management of national, or intra-EEZ, fishery resources, on the other hand, has lagged far behind such application to international fisheries management. This is now changing, as the strategic interaction among the relevant fishers, and between the fishers and the resource managers, can be ignored no longer. The application of game theory to the economic analysis of the management of national fisheries stands as the "new frontier".

This book is concerned with the relevance and application of game theory to the economics of the management of world capture (wild) fisheries. A game-theoretic situation is deemed to arise when the actions of one "individual" have a perceptible impact upon one or more other "individuals", leading to a strategic interaction between or among the "individuals". The relevance of the theory of such strategic interaction, popularly known as game theory, to the economics of capture fishery resource management has evolved gradually through time.

© Springer Nature Switzerland AG 2020
L. Grønbæk et al., *Game Theory and Fisheries Management*,
https://doi.org/10.1007/978-3-030-40112-2_1

1.1   Evolution of the Application of Game Theory to Fisheries Economics: The Era of Irrelevance

While modern fisheries economics can in fact be dated back to 1911 (Warming 1911), it received its fullest expression in 1954, with the publication of H. Scott Gordon's seminal article, "The economic theory of a common property resource: the fishery" (Gordon 1954). Over the two decades plus, following the publication of the Gordon article, game theory had no impact whatsoever upon the economics of capture fisheries management. Today, at the time of writing, game theory is to be seen as an indispensable tool in the development of the aforementioned economics.

The evolution of the relevance and application of game theory to the economics of capture fisheries management follows the evolution, although not precisely, of the economic management of capture fisheries. Stage I of such management can be referred to as no management. Capture fishery resources have been seen throughout history as the quintessential example of "common-pool" resources, because of the difficulty and costliness of establishing effective property rights to the resources. Up until the 1930s, the belief existed within the marine scientific community that, with the exception of a few isolated cases, the "common-pool" nature of these resources was not a matter for concern, because the resources were seen to be inexhaustible.[1] The inexhaustibility can be ascribed to economics and the then state of fisheries technology. It was too costly to exploit heavily fishery resources far from shore.

Stage II extends from the late 1930s until the late 1960s. Advances in fisheries technology, with a concomitant fall in harvesting costs, meant that the capture fishery resources were not inexhaustible after all. Evidence of overexploitation of these resources became evident to the scientific community by the late 1930s. The first response was to attempt to impose controls on season-by-season harvests, but with no control over fleet sizes. Given that under international law, at the time, coastal state jurisdiction over these resources did not extend beyond three nautical miles, most of the attempts had to be done internationally.

Gordon's (1954) article explores the economic consequences of an untrammeled "common-pool" fishery, one characterized by Pure Open Access.[2] The consequences are overexploitation of the fishery resource, misallocation of economic inputs and consequent dissipation of resource rent. This is contrasted with the fishery being under the control of a "sole owner". In the Pure Open Access case, it is assumed that the fishing industry is perfectly competitive (Gordon 1954). In either case, there is no strategic interaction, and thus no need for game theory.

If the fishery is subject to controls, but on the harvest only—no controls on the fleet size—we are then confronted with what is now commonly referred to as Regulated Open Access (Wilen 1985), which Gordon does in fact describe in his 1954 article (Gordon 1954). In any event, even though the overexploitation of the resource may

---

[1] Thus, for example, the great scientist of the nineteenth-century Britain, Thomas Huxley, stated in 1883 that the best fisheries management is no management, because of the inexhaustible nature of the fishery resources (cited in Bjørndal and Munro 2012, 7).

[2] Gordon assumes implicitly that the fishery resource is by no means inexhaustible.

be prevented, there will be an inevitable buildup of excess fleet capacity leading to the dissipation of resource rent. The assumption that the fishing industry is perfectly competitive is retained. Once again, there is no strategic interaction.

It is certainly true that attempts to impose harvesting controls at the international level involved strategic interaction among the states involved. This fact was essentially ignored. In part, this may be due to the fact that the international aspects of fisheries management were treated but lightly by economists at that time.[3] Furthermore, the argument has been made that, while the application of game theory to economics can be traced back to the mid-1940s, the widespread use of game theory by economists did not occur until the early 1970s (see, for example, Bierman and Fernandez 1993, 4). If the argument is valid, it would help to explain the lack of interest of fisheries economists in game theory. Be that as it may, we note that the highly influential book, *Mathematical Bioeconomics: The Optimal Management of Renewable Resources,* by Colin Clark, appeared first in 1976, and did so without a single reference to game theory Clark (1976). This stands in contrast to the second edition of the book, which appeared in 1990 (Clark 1990).

1.2 The UN Third Conference on the Law of the Sea and the Coming of the International Transboundary Fish Stock Issue: Game Theory Achieves Relevance

Stage III commences in the 1970s, with two events occurring simultaneously. The first, and by far the most important, was the UN Third Conference on the Law of the Sea, 1973–1982, leading to the 1982 UN Convention of the Law of the Sea (UN Convention, from hereon in) and the Exclusive Economic Zone (EEZ) regime (UN 1982). In spite of the difficulties of establishing property rights to capture fishery resources, the 1982 UN Convention enables coastal states to establish 200 nautical mile (370 km., approx.) EEZs. The fishery resources within a coastal state EZZ were and are to all intents and purposes coastal state property. It was estimated that, if the EEZ regime became universal, which is now close to being, the EEZs would encompass 90% of the world's commercially exploitable marine capture fishery resources (Bjørndal and Munro 2012, 9). There was thus a massive change in the status of capture fishery resources from international "common-pool" resources to coastal state property. This had two consequences. The first was a dramatic increase in the importance of capture fishery resource management at the national/regional level. The second consequence was the emergence of the international shared fish stock management problem.

---

[3] A prominent book of the time, which does, in fact, discuss in detail international attempts to control fishery exploitation, is *The Common Wealth in Ocean Fisheries*, by economists F. T. Christy and A.D. Scott. There is not even a hint in this discussion of the significance of strategic interaction among the states involved (Christy and Scott 1965, Chap. 11).

The second event, occurring at the national/regional level, was the attempt by many coastal states to address the Regulated Open Access problem by restricting the number of fishing vessels in individual fisheries, a management scheme commonly referred to as limited entry or licence limitation. From what has just been said, the coming of the EEZ regime greatly enhanced the significance of this change in fisheries management policy. With vessels entering a given fishery being limited by regulation, the assumption that the fishing industry is perfectly competitive ceases to be valid.[4] With the vessels in a fishery limited in number, strategic interaction among them becomes a distinct possibility.

It was, however, the emergence of the international shared fish stock management problem that led to the introduction of game theory to fisheries economics. We shall, as a consequence, focus initially on internationally shared fish stocks, and return to fisheries management at the national/regional level at a later point.

In any event, an internationally shared fish stock is defined as a stock that is exploited by two or more states. Prior to the coming of the EEZ regime, most capture fishery stocks were in fact internationally shared. As we have pointed out, the international aspects of fisheries management received very little attention from economists at that time. The coming of the EEZ regime brought the issue of internationally shared fishery resources to the fore. Coastal states establishing EEZs had no choice but to recognize that, because of the mobility of most fishery resources, some of the fish within the coastal state EEZ would almost certainly cross the border into neighbouring EEZs and/or would cross the border into the adjacent high seas, where they would be subject to exploitation by the so-called distant water fishing states (DWFSs).[5]

Employing FAO of the UN terminology, fish moving from one EEZ to another are defined as transboundary stocks, while those crossing the EEZ boundary into the adjacent high seas are defined as straddling stocks,[6] with the two *not* being mutually exclusive. To complete the terminology, those stocks confined to the remaining high seas are defined as discrete high seas stocks (Munro et al. 2004).

Two points are to be noted. While the UN Convention on the Law of the Sea was not available for signing before 1982, the fisheries issues in the Conference were settled by 1975. Coastal states began to plan accordingly, basing their plans

---

[4]The reader will recall that a fundamental assumption underlying the model of perfect competition is that there are no significant barriers of entry to or exit from the industry.

[5]A distant water fishing state is a fishing state, some of whose fleets operate well beyond domestic waters. While it is common to think of DWFSs as developed fishing state, e.g. Japan and Spain, there are developing fishing state DWFSs, as well, e.g. Ecuador, Mexico and Thailand. The typical DWFS is, of course, also a coastal state.

[6]The UN did, in fact, introduce another category, highly migratory stocks, e.g. tuna. These wide-ranging stocks do, by their very nature, cross the EEZ boundary into the adjacent high seas. The UN then defined straddling stocks as all other fish stocks crossing the EEZ boundary into the adjacent high seas. The distinction between the two arose out of bargaining in the UN Third Conference on the Law of the Sea. It has long been recognized that the distinction makes neither biological nor economic sense. This has led to the practice of combining the two into straddling stocks broadly defined (Munro et al. 2004).

upon drafts of the Convention.[7] Secondly, during the latter half of the 1970s and the early 1980s, straddling stocks and discrete high seas stocks were deemed to be unimportant. After all, it was estimated that the remaining high seas accounted for only 10% of the commercially exploitable marine capture fishery resources.

Thus, THE internationally shared fish stock issue of interest in the late 1970s and early 1980s was that of managing transboundary stocks, an issue, which coastal states were compelled to address. From the point of view of economists, the issue was tractable in that employing models involving but two coastal states were not unreasonable.

1.2.1 Transboundary Fish Stocks: The Legal Aspects

Before proceeding further, we must digress and consider the legal framework for the management of these stocks provided by the 1982 UN Convention on the Law of the Sea. The Convention has one provision pertaining to transboundary stock management, namely, Article 63(1). This article calls upon coastal states sharing such resources to seek to agree upon measures necessary to coordinate and ensure the conservation and development of such stocks, without prejudice to other provisions of the Convention (UN 1982, Article 63(1)).

Importantly, however, the 1982 UN Convention does not impose a duty upon the coastal states to reach an agreement. If the coastal states do not reach an agreement, then each coastal state is to manage the segment of the transboundary stock within its EEZ in accordance to the rights and duties relating to the management of intra-EEZ fishery resources as laid down by the Convention (Munro et al. 2004, 9). This can be referred to as the default option. Beyond all of this, the 1982 UN Convention has nothing further to say about the management of transboundary stocks.[8]

With the digression complete, let us continue. In 1977, one of the authors (Munro) undertook to prepare a conference paper on the economic management of transboundary fish stocks. Two questions were immediately seen to arise, with first being: what would the consequences be of coastal states sharing such a resource not cooperating in the management of the resource, i.e. the 1982 UN Convention default option? The second is: what conditions would have to be met for a cooperative resource management arrangement to be stable through time? A few papers, innocent of game theory, had been published on the issue, papers which Munro found to be incomprehensible. The incomprehensibility arose from the fact that the papers ignored the strategic

---

[7]The United States and Canada, for example, both established EEZs unilaterally in 1977.

[8]The UN Convention on the Law of the Sea, as noted, was obviously not finalized, until 1982. However, Article 63(1), as it appeared in the draft of the Convention available in the second half of the 1970s, was identical to that which appeared in the final version of the Convention.

interaction between/among the relevant coastal states, thereby ignoring the heart of the problem.[9]

Munro was thus compelled to face up to the realization that the then increasingly popular game theory would have to be brought to bear, whether he liked it or not.[10] It cannot be emphasized too strongly that this was not a case of a game theory specialist looking for an opportunity to display his/her skills. It was rather the case of a non-game theorist economist being forced by the policy issue at hand to turn to game theory, and to attempt to educate himself accordingly.

Munro turned to the second question, assuming that the coastal states, responding to the admonition of Article 63(1) of the 1982 UN Convention, are prepared to consider seriously the possibility of cooperative management of the shared fishery resource. In the chapters to follow, the reader will find him or herself being introduced to game theory concepts, and the types of game-theoretic models, in detail. Here we shall allow ourselves a broad-brush preview. The "individuals" engaged in strategic interaction, coastal states in our example, are referred to as players or agents. Their courses of action are referred to as strategies. The economic return to a player from following a particular strategy, given the reaction(s) of the other player(s), is referred to as the player's payoff. The implementation of the strategies by the players is the game, which may be static or dynamic. If there is an equilibrium outcome to the game, it is referred to as the solution to the game.

There are two very broad categories of games, competitive, non-cooperative, games and cooperative games.[11] If the players (agents) are playing competitively, we talk of them playing as "singletons"—a useful piece of jargon.

With respect to playing cooperatively, we do, in terms of the John Nash formulation (Nash 1953), think of the players cooperating for entirely selfish reasons only, in that they believe themselves to be better off by cooperating than by competing. This leads to two fundamental conditions that must be satisfied, if the solution to a cooperative game is to be stable, which we can refer to as the individual and collective rationality conditions. The solution must be individually rational in that each and

---

[9]Let a simple example suffice. Suppose that there are two coastal states, A and B, exploiting a transboundary fish stock. Except in unusual circumstances, the harvesting activities of the A fleet in the A EEZ will have an impact upon the harvesting opportunities open to the B fleet in the B EEZ, and vice versa—hence the strategic interaction.

[10]Munro happened to be very fortunate. At that very time, Munro's Department of Economics at the University of British Columbia (UBC) had a visitor from MIT, who presented a paper on optimal pricing policy for OPEC, with OPEC members having different views on such policy (Hnyilicza and Pindyck 1976). The visitor and his co-author employed John Nash's model of a two-player cooperative game (Nash 1953), largely ignored by economists up to that point. Munro asked himself if and how the game-theoretic analysis could be applied to the economics of transboundary fish stocks management. His colleague and co-author, Colin Clark of the UBC Department of Mathematics, provided invaluable assistance. By further chance, the UBC Department of Mathematics had in residence at that time a prominent visitor from Paris, who was a specialist in game theory.

[11]Having said this, Chap. 2 will make apparent that the distinction is in fact not clear-cut. While one can think of non-cooperative games that are strictly non-cooperative, cooperative games have non-cooperative games lurking in the background. Indeed, the alert reader will detect this in the discussion to follow.

every player, at each and every moment in time, must be assured of a payoff at least as great as that which the player would enjoy under competition. The solution must be collectively rational in that it is Pareto optimal; there cannot exist an alternative solution, which would make one player better off, without harming the other(s). The conditions appeal to common sense, although it must be remarked that, in the real world of cooperative management of international fishery resources, common sense is often found to be uncommon.

In the case of fishery cooperative games, there is a third condition, a dynamic condition. For want of a better term, we can refer to it as the resiliency condition. In the cooperative management a fishery resource, allowance must be made for the distinct possibility that conditions will change over time, due to environmental, economic or other factors. It may be possible to anticipate such changes, but they cannot, in most instances, be predicted with any degree of accuracy. If the solution to the fishery cooperative game is to be stable through time, the cooperative resource management arrangement must be able to withstand unpredictable shocks. In other words, the cooperative resource management arrangement must be *resilient*. Once again, the condition upon being stated is obvious. In the real world, the obvious is often ignored.

Munro's cooperative fishery game model assumes only two players (looking to Canada–US examples) and assumes that the players are asymmetric, allowing them to differ in numerous ways, e.g. social rates of discount, harvesting costs (Munro 1978). The conference paper was revised and extended, with the revised paper appearing as journal article in 1979 (Munro 1979).

The paper/article attempts to demonstrate that, even with asymmetries leading to differences in perceived optimal resource management policies, stable solutions to cooperative fisheries management games are obtainable. In his education and preparation of the paper, Munro was introduced to the concept of side payments—transfers—between/among players, which can be monetary or non-monetary in form.[12] He discovered that bringing in side payments led to a massive simplification of the analysis, which led him in turn to suspect that side payments are not without relevance to the world of policy.[13]

It was stated that the economic management of transboundary fishery resources raises two questions. Munro had not addressed the first one, the consequences of the coastal states sharing a transboundary fishery resource not cooperating. If the negative consequences are negligible, then there is not much point cooperating, and Munro's analysis would be beside the point. The first question was addressed in

---

[12]In later chapters, side payments will be referred to as transferable utility.

[13]It also led him to advance what he referred to as the *Compensation Principle* (Munro 1987). Consider a two-player fishery cooperative game, where, due to the aforementioned asymmetries, the players have different views on the optimal resource management policy. The *Compensation Principle* states that, in the case described, it will almost invariably be found that one player places a greater value on the resource than does the other. Optimal management calls for the player placing the greater value on the resource to dominate the resource management policies. In order for all of this to be feasible, that player must be prepared to compensate the other player through side payments.

two articles appearing soon thereafter in 1980, by Clark (1980) and by Levhari and Mirman (1980). The two articles are in agreement.

The game theory analysis relevant to the first question is obviously the theory of competitive (non-cooperative) games. The reader will be introduced to this theory in Chap. 2, and explore this theory in detail in Chap. 3. In so doing, the reader will become acquainted with the Prisoner's Dilemma (re-named the Fisher's Dilemma). The point of the Prisoner's Dilemma is that in a competitive game the players can easily be driven to adopt strategies, which they recognize to be harmful. The Prisoner's Dilemma is prominent in both articles. Clark, for example, demonstrates, in the context of a two-player dynamic competitive game, that the players can be led to drive the transboundary resource down to a common Bionomic Equilibrium, which we associate with Pure Open Access (Clark 1980). The conclusion arising from both Clark (1980) and Levhari and Mirman (1980) is that, except in unusual circumstances, non-cooperative management of a transboundary fishery resource will result in inferior resource management. Cooperation does matter.

The Munro analysis of 1979 suffers from two major limitations. First, the analysis assumes that, if the players establish a cooperative resource management agreement, the agreement will be binding over time, even though conditions may change through time. This is unrealistic. Even though the agreement may take the form of a formal treaty, a dissatisfied player can often find means of making the treaty unworkable. The second limitation is the assumption that the players are but two. Even in the case of transboundary fish stocks, there are numerous cases in which the number of players exceeds two by a wide margin.

The first limitation was addressed initially in the 1980s, and later in the following decade, with models in which it is assumed that cooperative agreements, upon being reached, are not strictly binding over time (e.g. Kaitala and Pohjola 1988). Basically, what then becomes necessary are effective punishment schemes sufficient to deter would-be defectors, schemes which will be discussed further in Chap. 7.

The second limitation is more far-reaching, and came to be addressed in the 1990s (e.g. Kaitala and Lindroos 1998). Once the number of players exceeds two, then we must allow for the possibility of sub-coalitions being formed, which compels us to enter the realm of coalition games. An important issue, which arises, is that of optimal rules for the sharing of the net economic benefits arising from the cooperative management. Optimal sharing is straightforward when the players are only two, as will be seen in Chap. 4. When the number of players exceeds two, optimal sharing becomes a matter of considerable complexity, to be addressed through the application of what are termed characteristic function games. All of these issues will be explored at length in Chap. 5.[14]

---

[14] Another, but lesser, limitation of the Munro analysis is that it relies entirely upon the Nash model of cooperative games. There are other valid alternatives, a point made forcefully by Armstrong (1994).

1.3   The Emerging International Straddling Fish Stock Issue and the Need for Game-Theoretic Models of Increasing Complexity

Stage IV in the management of world capture fishery resources occurred in the 1980s and the 1990s. There were developments at both the international and national/regional levels. We postpone our discussion of the developments at the national/regional level.

At the international level, the comfortable assumption that straddling stocks (broadly defined) could be ignored was proven to be invalid. Case after case of straddling fish stocks being the basis of serious resource mismanagement arose. By the early 1990s, the UN felt compelled to deal with the issue, and did so by convening a conference, popularly known as the UN Fish Stocks Conference,[15] 1993–1995. The Conference brought forth an agreement, popularly known as the UN Fish Stocks Agreement (UN 1995),[16] which serves to supplement the 1982 UN Convention (Munro et al. 2004). We must examine the Agreement in some detail, with yet another digression.

1.3.1   Straddling Fish Stocks: The Legal Aspects

The legal aspects of the management of straddling fish stocks are far more complex than those of transboundary fish stocks. To begin, we must recall the complexity introduced by the fact that these fishery resources are exploited both by coastal states and by distant water fishing states (DWFSs).[17]

We must next note that, going back to the UN Third Conference on the Law of the Sea, the UN faced a delicate problem of balance. It had to balance the rights of coastal states in managing fishery resources within the EEZs against the Freedom of the Sea rights of DWFSs operating in the remaining high seas. The attempt to achieve such a balance through the 1982 UN Convention was unsuccessful, as evidenced by the growing straddling fish stocks problem (Munro et al. 2004). The consequence was the UN Fish Stocks Conference and the resultant 1995 Agreement. The purpose of the Agreement, as the full title of the Agreement makes clear, is not to replace any part of the 1982 UN Convention. It is rather to buttress the Convention.

The purpose of the Agreement is to bring together relevant coastal states and DWFSs for the cooperative management of the high seas segments of straddling stocks. Such management, however, must be compatible with that of the intra-EEZ

---

[15]UN Conference on Straddling Fish Stocks and Highly Migratory Fish Stocks. See no. 6.

[16]The full title of the Agreement is: Agreement for the Implementation of the Provisions of the United Nations Convention on the Law of the Sea of 10 December 1982 Relating to the Conservation and Management of Straddling Fish Stocks and Highly Migratory Fish Stocks.

[17]See no. 5.

segments of the stocks.[18] The duty of these states to cooperate is much stronger than it is in the case of transboundary stocks (Munro et al. 2004).

The mechanism preferred by the Agreement for achieving this cooperation consists of what are called Regional Fisheries Management Organizations (RFMOs). Examples are provided by the Northwest Atlantic Fisheries Organization (NAFO), off Canada and the United States, and the Western and Central Pacific Fisheries Commission (WCPFC).

All states having "real interest" in the relevant fisheries covered by an RFMO have the right to participate in an RFMO (UN 1995, Article 8(3)). The definition of "real interest" is, however, vague, which is a matter of importance, since the Agreement states that only those states, which are members of the RFMO, or agree to abide by the management provisions of the RFMO, are to have access to the fishery resources covered by the RFMO (UN 1995, Article 8 (4)).

Furthermore, a distinction has to be made between what one might call the "charter" members of an RFMO, and states which appear on the scene at a later date—"new members". A "new member", expressing real interest in the relevant fishery resources, is not granted automatic membership. The Agreement does indeed make it clear that existing members of an RFMO must be prepared to accommodate "new members". The Agreement also, however, calls upon existing members to take into account various considerations upon receiving applications from "new members", with one such consideration being the status of the relevant stocks and the existing level of fishing effort in the fisheries (UN 1995, Articles 8, 10 and 11). Thus, the existing members of an RFMO may say to a prospective "new member" that they would be delighted to accept the "new member" into the club, but that alas the relevant fisheries are "full up" (Munro et al. 2004).

Added to these complications is the fact that, under international treaty law, a treaty is binding only upon those states, which have ratified the treaty,[19] unless the treaty has gained the status of "customary international law".[20] One consequence is that there is some degree of ambiguity as to the property rights to the fishery resource in the high seas under the jurisdiction of an RFMO. Fishing vessels, which are operating in these high seas in a manner contrary to the management provisions of the RFMO, are deemed to be engaging, not in illegal fishing, but rather in unregulated fishing (Munro et al. ibid.)

The overall consequence of all of this is that the management of straddling fish stocks through RFMOs is a far, far more difficult problem than that of managing strictly transboundary fish stocks.[21] To begin, the typical RFMO has a large number

---

[18] Which means, in effect, that the management of the entire stocks must be addressed.

[19] *Pacta tertiis nec nocent nec prosunt.*

[20] Customary international treaty law is treaty law, which is so widely accepted and practiced, that states, which have not ratified the treaty, are bound to accept the treaty's provisions, unless they state explicitly their unwillingness to do so (Buergenthal and Murphy 2002). It is not clear to these authors whether the Agreement has, or has not, achieved the status of customary international treaty law.

[21] We refer to strictly transboundary fish stocks, because, as it will be recalled, a fish stock can be both transboundary and straddling in nature.

of members (players). Game theory models with three or four players, let alone two, are of very limited value. Hannesson (1997) was one of the first to address this large number issue.

The large number of players in an RFMO game, combined with the ambiguity of fish resource property rights in the high seas under the jurisdiction of RFMOs, aggravated by the "new member" problem, has led to the result that free riding is a chronic problem facing the typical RFMO.[22] The stability of RFMOs through time becomes the key question.

RFMOs vary greatly in their ability to deal with unregulated fishing—free riding. To begin, the problem of such free riding is now widely recognized by policymakers (Munro 2013). Several RFMOs have been vigorous in their attempts to stamp out unregulated fishing, to the extent that such RFMOs have been cooperating with one another in what might be termed an informal marine Interpol. Other RFMOs have been much less vigorous (Munro 2013).

The game-theoretic models that have proven to be most effective in addressing this question consist of partition function games, introduced to the fisheries economics literature by one of the authors (Pintassilgo 2003). Partition function games, the "new member" problem, and the stability of RFMOs over the long run will be examined in detail in Chap. 6.

1.3.2 Impact of Game-Theoretic Analysis of International Fisheries Management Upon Policymakers

A key, indeed fundamental, question arises, which we need to address before proceeding further. Has the application of game theory to the economics of the management of internationally shared fish stocks had any impact upon policymakers, or has it been of interest to academics alone?

The basic answer to the question is that the game-theoretic analysis has indeed had an impact upon policymakers, but that the impact did not occur without a significant lag. In terms of policy pertaining to the management of international fisheries, the FAO of the UN plays a central role. By the mid to late 1990s, there were clear indications that the results of the game-theoretic analysis were penetrating the FAO. An example is provided by a then prominent and senior member of the Fishery Resources Division of FAO, John F. Caddy. Caddy, who is a marine biologist by training, but who developed a strong interest in the economics of fishery management, was called upon in the mid-1990s to deliver a paper at a conference in Australia, with the conference title being: Taking Stock: Defining and Managing Shared Resources.

Caddy's paper is focused on the management of transboundary fish stocks (Caddy 1997). In the paper, he states that "—it is not surprising that game theory, widely

---

[22]In theory, free riding could also be a problem in the economic management of transboundary fish stocks. Nonetheless, the evidence indicates that free riding is, in fact, not a major issue in the management of such stocks (Munro et al. 2004).

employed in economics, military strategy and any other transactions involving two or more parties should begin to find application in fishery negotiations" (Caddy 1997, 101). What then follows is a lengthy section, citing numerous sources, on the application of game theory to the analysis of the management of transboundary fish stocks.[23]

Five years after the publication of the Caddy paper, the FAO, in cooperation with the government of Norway, mounted an Expert Consultation on the Management of Shared Fish Stocks (FAO 2002). In the report on the Consultation, the following appears in the summary of the main conclusions arising from the Consultation. After observing that "the management of shared fishery resources remains as one of the great challenges on the way to achieving long-term sustainable fisheries" (FAO 2002, iv), the report puts forth as the first conclusion of the Consultation that, with few exceptions, non-cooperative management of shared fish stocks carries with it the threat of overexploitation. Cooperation in the management of these resources is an essential pre-requisite, if the management is to be effective.

The next conclusion is that no attempt to achieve cooperative management of shared stocks can hope to succeed "unless each and every participant anticipates receiving long-term benefits from the cooperatively managed fishery that are at least equal to the long-term benefits it would expect to receive in the absence of collaboration" (FAO 2002, ibid.). The Consultation notes that this fact, while apparently obvious, is often ignored.

The Consultation goes on to conclude that, in order for cooperative management arrangements to succeed through time, the arrangements must be resilient, being able to absorb unpredictable shocks. The Consultation next concludes that recognition must be given to the importance of side payments, with the Consultation preferring the euphemism, "negotiation facilitators". These negotiation facilitators, argues the Consultation, would "—broaden the scope for bargaining over allocations, assist in achieving compromises when there are differences in the management goals of cooperating States/entities, and enhance the flexibility and resilience the cooperative arrangements over time" (FAO 2002, ibid.).

The Consultation led to the preparation of a FAO Fisheries Technical Paper on the legal and economic aspects of the conservation and management of shared fish stocks, designed, *inter alia*, for policymakers. The Technical Paper explicitly introduces a review of the basic concepts of competitive and cooperative game theory, and then proceeds to employ the concepts (Munro et al. 2004).[24]

A few years after the FAO Expert Consultation, there was established, as the result of an OECD initiative, an Independent Panel to Develop a Model for Improved Governance of Regional Fisheries Management Organizations, based at the Royal Institute of International Affairs (Chatham House) in London. The Panel brought down its report in 2007 (Lodge et al. 2007). In the summary and overview, the following appears: "—a core conclusion is that the success of international cooperation depends largely on the ability to deter free-riding. When few countries exploit a

---

[23]Included in which there is a detailed exposition on the *Compensation Principle*. See no. 13.

[24]See in particular: Munro et al. (2004), Parts 3.2.1–3.2.4, inclusive.

resource, free-riding can be deterred. When the number of countries is large, however, free-riding is much more difficult to prevent—. A good RFMO will (a) require that more be done to conserve and manage the stocks at an optimum level—(b) create incentives for States to participate, and (c) create incentives for parties to comply" (Lodge et al. 2007, x).

A more recent development occurred in 2019. With the UK as a member of the EU, the involvement of the UK fishing industry and government in international fisheries negotiations has been indirect through the EU. If Brexit comes to pass, then the UK will be on its own, and indeed will have to enter international fisheries negotiations with its former EU partners, as well as others. In preparation for this eventuality, the UK Sea Fish Industry Authority (Seafish) mounted a one-day intensive workshop for senior representatives from UK fishing industry and government at Chatham House in London, in early May 2019. The title of the workshop: "Economics and Game Theory for International Fisheries Agreements" (Seafish 2019).[25]

1.4 The Management of Fish Stocks at the National/Regional Level

We come at last to the relevance of game theory to economic management of fisheries at the national/regional level. Game theory is becoming increasingly relevant at this level, but the application of game-theoretic analysis at this level lags far, far behind the application of the analysis to economic management of international fisheries.

If we return to the stages of the economic management of fisheries, it will be recalled that in Stage II, the management consisted of what has been termed Regulated Open Access. Attempts were made to control global harvests, thereby conserving the resources, but no attempts were made to control the fleet sizes. The result was the buildup of excess capacity and the dissipation of resource rent. It is a straightforward matter to construct an economic model of the fishery predicting this outcome, and to do so without the use of game theory (see, for example, Munro and Scott 1985, 613).

We stated further that in Stage III, commencing in the early 1970s, there was at the national/regional level, in many coastal states, an attempt made to address the Regulated Open Access problem by restricting the number of vessels in individual fisheries—limited entry or licence limitation. The assumption that the fishing industry is perfectly competitive ceases to be valid, with strategic interaction among the vessel owners becoming a distinct possibility.

If a strategic interaction among vessel owners does occur, then we have to think not only in terms of a fisher (vessel owner) game, but also in terms of a game between the industry and the resource manager. This leads us into what are known as stage games. The appropriate framework was first formally set out by Kronbak and Lindroos (2006). In their model, the resource manager is seen as playing a leader–follower

---

[25]Two of the authors of this book were invited to make presentations at the workshop.

game with the industry. There is in turn an intra-industry game. At the first stage, the resource manager makes its move by setting out fishery management regulations. In the following stages, the industry reacts. The industry game may be competitive, partially cooperative or fully cooperative. Analysis reveals, to no surprise, that the resource manager is better off, the greater is the degree of cooperation among the industry members (Kronbak and Lindroos 2006).[26]

In the Stage III limited entry programmes, the fisher game was strictly competitive. Indeed, the regulations were such that the fishers were encouraged to compete for shares of the limited season-by-season global harvest (Bjørndal and Munro 2012). Wilen (1985) explored the consequences, employing the theory of competitive games.

The objective of the resource managers is to control and limit fleet capacity (the ability to catch fish). Controlling the number of vessels is inadequate, because capacity is commonly made up of a set of substitutable inputs. Resource managers have in practice found it virtually impossible to control all inputs.

Wilen demonstrates that, even if all vessel owners realize that their attempts to expand fishing capacity will dissipate resource rents, each vessel owner has no choice but to attempt to expand its capacity. In other words, we are presented with a textbook example of the Prisoner's Dilemma (Wilen 1985). The Wilen predictions have been validated over and over. In fact, in many instances, not only has resource rent been dissipated, but the resource manager's ability to regulate effectively the season-by-season global harvest has been undermined (Clark and Munro 2017).

We said earlier that there was a Stage IV in the economic management of world capture fisheries, occurring in the 1980 and 1990s, with developments at both the international and national/regional levels. The discussion of the developments at the national/regional level were postponed. That discussion can be postponed no further.

At Stage IV, there was, at the national/regional level, a reaction to growing evidence of the inadequacies of limited entry type programmes. From an economic standpoint, the problem with common-pool fisheries has been seen to be the perverse incentives created for fishers, leading to overexploitation of the resources and/or the buildup of excess fleet capacity. This has led the FAO to talk in terms of Incentive Blocking and Incentive Adjusting approaches to fisheries management, with the former referring to measures to block fishers from responding to the perverse incentives, and the latter referring to measures to adjust fisher incentives, such as to bring them more closely in line with social objectives (Bjørndal and Munro 2012). Limits on global harvests, gear restrictions and limits on entry to individual fisheries are all examples of Incentive Blocking measures.

The aforementioned reaction in Stage IV took the form of turning to Incentive Adjusting approaches, but without a wholesale abandonment of Incentive Blocking measures. The Incentive Adjusting approaches took two forms, taxes and harvesting rights-based fisheries schemes, popularly referred to as "catch share" schemes. Taxes, positive or negative, constitute the economist's classic measure for adjusting incentives in a very broad set of circumstances. In the case of fisheries, the use of taxes has been analysed extensively by economists, but applied only sparingly.

---

[26]The Kronbak and Lindroos (2006) model will be discussed in greater detail in Chap. 7.

In terms of application, what has been most widely used in the Incentive Adjusting category takes the form of "catch shares". The basic idea is that, instead of having fishers compete for shares of the total allowable harvest, the total allowable harvest is divided up among the fishers. The hope is that this will eliminate or at least mitigate the fishers' incentive to expand capacity. The shares may be allocated to fishers on an individual basis, IQs, individual quotas, or on a collective basis, leading to the establishment of fisher cooperatives.[27] The most prominent (at least among developed fishing states) are IQs, now more commonly ITQs, individual transferable quotas.

To the extent that taxes and IQs have been studied using game-theoretic analysis, the usual assumption is that the fishers are playing as singletons. This is eminently sensible in the case of taxes, but less so, as we shall argue, in the case of IQs. The most common form of analysis applied takes the form of Principal–Agent analysis, where the resource manager is seen as the Principal and the fishers as the Agents (e.g. Vestergaard 2010). A Principal–Agent situation can be seen as being essentially the same as a leader–follower game (Mesterton-Gibbons 1993).[28]

If the fishers are playing as singletons in a tax or catch share scheme, the tax or catch share schemes may mitigate the consequences of fisher competition, but they do not entirely eliminate those consequences. As Ragnar Arnason states in his review of ITQ schemes, in which the fishers are playing as singletons, ITQs carry us far towards optimality, but not far enough. Many inefficiencies persist (Arnason 2017).

Kronbak and Lindroos (2006), it will be recalled, put forward the possibility of fishers playing cooperatively, with resultant positive benefits for the resource manager and society at large. Is it not possible that Incentive Adjusting schemes could be designed in such a way as to foster intra-industry cooperation among the fishers? With respect to tax schemes, the answer is presumably no. Catch share schemes are a different matter.

To begin, successful fisher cooperatives are, by definition, stable cooperative games. On the other hand, it was thought until recently that the scope for cooperation among ITQ holders is very limited (e.g. Arnason 2012). This, however, has now been disproven. There exists a widely documented case of an ITQed fishery off Pacific Canada, in which the "players", while large in number, are, and have been, playing a stable cooperative game (Wallace et al. 2015; Grønbæk et al. 2016; Clark and Munro 2017).

Furthermore, the Pacific Canada case reveals a development not considered by Kronbak and Lindroos (2006). In the Pacific Canada case, the industry–resource manager game was, in the past, a highly competitive one. That game, for reasons not entirely clear, has evolved into a cooperative one. This development, in turn, has given rise to de facto co-management. While the ultimate resource management decisions rest with the resource manager, the fishers have been able to influence

---

[27]There are also schemes referred to as community-based fisheries management, or territorial use rights fisheries (TURFs). These are best thought of as fisher cooperatives having a specific geographical location.

[28]Principal-Agent analysis in fisheries is a further topic to be discussed in Chap. 7.

resource management and have done so in a positive direction. There are recorded instances in this fishery in which the fishers have pressed the resource manager to slash TACs on certain stocks, with the fishers suspecting before the resource manager that the stocks were endangered. The followers have turned leaders, proving to be more conservationist than the resource manager (Grønbæk et al. 2016).[29]

How many other cases similar to that of the Pacific Canada fishery exist throughout the world, involving not just ITQed fisheries, but also fisher cooperatives? This we do not know.

If these cases, in which the fishers are playing cooperatively among themselves, and then in turn playing cooperatively with the resource manager, were to become widespread, the implications for the resource management at the national/regional level would be profound. We could be said to have entered a new stage of the economic management of capture fishery resources, if not Stage V, then at least Stage IV+.

What game-theoretic analysis then has been brought to bear setting forth the conditions that must prevail for this double level cooperation to be stable through time? The answer is little, very little indeed. This stands as one of the important areas of future game-theoretic research in fisheries economics. All of this will be discussed further, and at length, in Chap. 7, and again in Chap. 8.

References

Armstrong, C. (1994). Co-operative solutions in a transboundary fishery: The Russian—Norwegian co-management of the Arcto-Norwegian cod stock. *Marine Resource Economics, 9*(4), 329–351.

Arnason, R. (2012). Property rights in fisheries: How much can individual transferable quotas accomplish? *Review of Environmental Economics and Policy, 6,* 217–236.

Arnason, R. (2017). *Catch shares: Potential for optimal use of marine resources.* Presentation to the North American Association of Fisheries Economists Forum 2017, La Paz, Mexico, 22–24 March 2017.

Bierman, H., & Fernandez, L. (1993). *Game theory with economics applications.* New York: Addison-Wesley.

Bjørndal, T., & Munro, G. (2012). *The economics and management of world fisheries.* Oxford: Oxford University Press.

Buergenthal, T., & Murphy, S. (2002). *Public international law.* St. Paul: West Group.

Caddy, J. (1997). Establishing a consultative mechanism or arrangement for managing shared stocks within the jurisdiction of contiguous states. In D. Hancock (Ed.), *Taking stock: Defining and managing shared resources.* Australian Society for Fish Biology and Aquatic Resource Management Association of Australasia Joint Workshop Proceedings, Darwin, NT, 15–16 June 1997 (pp. 81–223). Sydney: Australian Society for Fish Biology.

Christy, F., & Scott, A. (1965). *The common wealth in ocean fisheries.* Baltimore: Johns Hopkins University Press.

Clark, C. (1976). *Mathematical bioeonomics: The optimal management of renewable resources* (1st ed.), New York: Wiley-Interscience.

Clark, C. (1990). *Mathematical bioeonomics: The optimal management of renewable resources* (2nd ed.), New York: Wiley-Interscience.

---

[29]The Pacific Canada fishery in question will be reviewed in detail in a case study in Chap. 7.

Clark, C. (1980). Restricted access to a common property resource. In P. Liu (Ed.), *Dynamic optimization and mathematical economics* (pp. 117–132). New York: Wiley-Interscience.

Clark, C., & Munro, G. (2017). Capital theory and the economics of fisheries: Implications for policy. *Marine Resource Economics, 32*(2), 123–142.

Food and Agriculture Organization of the United Nations. (2002). *Report of the FAO—Norway expert consultation on the management of shared fish stocks*, Bergen, Norway, 7–10 October 2002, FAO Fisheries Report No. 695, Rome.

Gordon, H. (1954). The economic theory of a common property resource: The fishery. *Journal of Political Economy, 62,* 124–142.

Grønbæk, L., Lindroos, M., Munro, G., & Turris, B. (2016). *Application of game theory to intra-EEZ fisheries management*. Paper prepared for the 18th Biennial Conference of the International Institute of Fisheries Economics and Trade, Aberdeen, Scotland, July 2016.

Hannesson, R. (1997). Fishing as a supergame. *Journal of Environmental Economics and Management, 32,* 309–322.

Hnyilicza, E., & Pindyck, R. (1976). Pricing polices for a two-part exhaustible resource cartel: The case of OPEC. *European Economic Review, 8,* 139–154.

Kaitala, V., & Pohjola, M. (1988). Optimal recovery of a shared resource stock: A differential game with efficient memory equilibria. *Natural Resource Modeling, 3,* 91–117.

Kaitala, V., & Lindroos, M. (1998). Sharing the benefits of cooperation in high seas fisheries: A characteristic game approach. *Natural Resource Modeling, 11,* 275–299.

Kronbak, L., & Lindroos, M. (2006). An enforcement-coalition model: Fishermen and authorities forming coalitions. *Environmental & Resource Economics, 35,* 169–194.

Levhari, D., & Mirman, L. (1980). The great fish war: An example using a dynamic Cournot-Nash solution. *Bell Journal of Economics, 11,* 649–661.

Lodge, M., Anderson, D., Løbach, T., Munro, G., Sainsburyt, K., & Willock, A. (2007). *Recommended best practices for regional fisheries management organizations: Report of an independent panel to develop a model for improved governance by regional fisheries management organizations*. London: Chatham House.

Mesterton-Gibbons, M. (1993). Game-theoretic resource modeling. *Natural Resource Modeling, 7,* 93–147.

Munro, G. (1978). Canada and extended fisheries jurisdiction in the Northeast Pacific: Some issues in optimal resource management. *Proceeding of the 5th Pacific Regional Science Conference*, Bellingham, Western Washington University (pp. 3–16).

Munro, G. (1979). The optimal management of transboundary renewable resources. *Canadian Journal of Economics, 12,* 355–376.

Munro, G. (1987). The management of shared fishery resources under extended jurisdiction. *Marine Resource Economics, 3,* 271–296.

Munro, G. (2013). Regional fisheries management organizations and the new member problem: From theory to practice. In A. Dinar & A. Rapoport (Eds.), *Analyzing global environmental issues: Theoretical and experimental applications and their policy applications* (pp. 106–128). New York: Routledge.

Munro, G., & Scott, A. (1985). The economics of fishery management. In A. Kneese & J. Sweeney (Eds.), *Handbook of natural resource and energy economics* (Vol. II, pp. 623–676), Amsterdam, North-Holland.

Munro, G., Van Houtte, A., & Willmann, R. (2004). *The conservation and management of shared fish stocks: Legal and economic aspects*. Food and Agriculture Organization of the UN, Fisheries Technical Paper, No. 465, FAO, Rome.

Nash, J. (1953). Two-person cooperative games. *Econometrica, 21,* 128–140.

Pintassilgo, P. (2003). A coalition game approach to the management of high seas fisheries in the presence of externalities. *Natural Resource Modeling, 16*(2), 175–192.

Seafish. (2019). Retrieved from https://www.youtube.com/watch?v=BX14AStbRyk.

United Nations (1982). *United Nations convention on the law of the sea*. UN Doc. A/Conf. 62/122.

United Nations. (1995). United Nations Conference on Straddling Fish Stocks and Highly Migratory Fish Stocks. Agreement of the Implementation of the Provisions of the United Nations Convention on the Law of the Sea of 10 December 1982 Relating to the Conservation and Management of Straddling Fish Stocks and Highly Migratory Fish Stocks. UN Doc. A/ Conf./164/37.

Vestergaard, N. (2010). Principal—agent problems in fisheries. In R. Grafton, R. Hilborn, D. Squires, M. Tait, & M. Williams (Eds.), *Handbook of marine fisheries conservation and management* (pp. 563–571). Oxford: Oxford University Press.

Wallace, S., Turris, B., Driscoll, J., Bodtker, K., Mose, B., & Munro, G. (2015). Canada's Pacific groundfish trawl habitat agreement: A global first in an ecosystem approach to bottom trawl impacts. *Marine Policy, 60,* 240–248.

Warming, J. (1911). Om grundrente af fiskegrunde. *Nationaløkonomisk Tidsskrift, 49,* 499–505.

Wilen, J. (1985). Towards a theory of the regulated fishery. *Marine Resource Economics, 1,* 69–88.

Chapter 2
Basic Concepts in Game Theory

Abstract This chapter defines the key concepts to understand a fishery game. In particular, the concepts of game, strategy, and the representation of a game in strategic and extensive forms. The most widely used solution concept in game theory, the Nash equilibrium, is presented as well as arguments that support its applicability. The two basic approaches in game theory, non-cooperative games and cooperative games, are outlined. Finally, the North Sea herring fishery is presented as an illustrative case study.

In Chap. 1, it was said that a game-theoretic situation emerges when the actions of a decision-maker have a perceptible impact upon one or more decision-makers, and vice versa, leading to a strategic interaction between or among them. The theory of strategic interaction, or interactive decision theory, popularly known as game theory, is designed to analyse the behaviour of these decision-makers, as they interact with one another. Chapter 1 went on to state that the decision-makers interacting with one another are referred to as "players", while their courses of action are termed strategies. The economic return to a player following a particular strategy, given the reaction(s) of the other player(s), is referred to, in turn, as the player's payoff. The implementation of the strategies by the players is the game, which may be non-cooperative or cooperative. Finally, if there is an equilibrium outcome to the game, it is referred to as the solution to the game.

It is now necessary that we move beyond these useful, but broad-brush, definitions to definitions that are both precise and rigorous. This is the primary purpose of this chapter.

2.1 What is a Game?

Game theory is anchored in the concept of game, which can be defined as follows (Mas-Collel et al. 2012).

© Springer Nature Switzerland AG 2020
L. Grønbæk et al., *Game Theory and Fisheries Management*,
https://doi.org/10.1007/978-3-030-40112-2_2

> **Definition 2.1** A game is a formal representation of a situation in which a number of decision-makers interact in a setting of strategic interdependency.

The formal representation takes the form of a mathematical model, which involves four elements (Mas-Collel et al. 2012):

(i)   The players: the decision-makers.
(ii) The rules of the game: Who moves when? What do they know when they move? What can they do?
(iii) The outcomes: For each possible set of actions of the players, what is the outcome of the game?
(iv) The payoffs: the players' utilities over the possible outcomes.

In a game, each player will set strategies.

> **Definition 2.2** A strategy is a complete contingent plan, or decision rule, that specifies how a player will act on each possible distinguishable circumstance in which she might be called upon to move (Mas-Collel et al. 2012).

Consider a n-player game and let the set of players be represented by $N = \{1, 2, \ldots, n\}$. The strategy of each player $i \in N$ can be represented by s_i and the profile of strategies of all players by the vector $s = (s_1, s_2, \ldots, s_n)$. Each profile of strategies will generate an outcome of the game and consequently the payoff for each player. From this direct link between strategies and payoffs emerges the strategic (or normal form) of representing a game.

> **Definition 2.3** For a game with n-players, the **strategic (or normal)** representation Γ specifies for each player i, a set of strategies S_i (with $s_i \in S_i$) and a payoff function $u_i(s_1, s_2, \ldots, s_n)$ given the von Neumann–Morgenstern utility levels associated with the (possible random) outcome arising from strategies $(s_1, s_2, \ldots, s_n)$. Formally, we write $\Gamma = [N, \{S_i\}, \{u_i(.)\}]$ (Mas-Collel et al. 2012).

Therefore, as referred by Fudenberg and Tirole (1991), a game in strategic (or normal) form is completely defined by the set of players, $N = \{1, 2, \ldots, n\}$, the set of strategies of each player i, S_i, and the payoffs $u_i(s)$ for each profile of strategies, $s = (s_1, s_2, \ldots, s_n)$. The following example illustrates the main concepts of a game.

Example 2.1 Consider a fish stock harvested by two symmetric countries A and B (the players), that is, both countries share the same technology, harvesting costs and fish price. Assume that the two countries decide simultaneously what level of constant fishing effort, in terms of number of boats, to apply in the fishery (rules of the game). Each country will either apply their optimal fishing effort under non-cooperation (NC: 20 boats) or the optimal fishing effort under cooperation (C: 10 boats). Thus, there are four possible outcomes of the game: (C, C); (C, NC); (NC, C); (NC, NC), where the first cell of each vector represents the action of country A, and the second cell the action of country B. Each country has four possible strategies. For instance, the strategies of player A are as follows:

Strategy 1 $\left(s_1^1\right)$: Play C if player B plays C; Play C if player B plays NC.
Strategy 2 $\left(s_1^2\right)$: Play C if player B plays C; Play NC if player B plays NC.
Strategy 3 $\left(s_1^3\right)$: Play NC if player B plays C; Play C if player B plays NC.
Strategy 4 $\left(s_1^4\right)$: Play NC if player B plays C; Play NC if player B plays NC.

The strategic (or normal) form of the game is shown in Fig. 2.1.

The values in Fig. 2.1 represent the payoffs of the players, which in this example can be measured in monetary units. When both players choose the cooperative fishing effort (C: 10 boats), they earn a payoff of 4. If the outcome is (NC: 20 boats, C: 10 boats), that is, player A chooses the non-cooperative fishing effort level and player B the cooperative level, then the payoff of player A will be 5, exceeding its payoff when both countries adopt the cooperative level. In this outcome, player A would free ride the conservative efforts of player B, who would get the lowest possible payoff in the game (0). If both players choose the non-cooperative fishing effort level, then their payoffs (1) will be significantly lower than what they would earn if both adopted a cooperative level (4).

An alternative way to represent a game is called the extensive form. It is a detailed description of the sequential structure of the decision problems of the players based on a game tree (Osborne and Rubinstein 2016). The tree shows who moves when, the actions each player can take, what players know when they move, the outcomes of the game and the players' payoffs from each possible outcome (Mas-Collel et al. 2012). An extensive form of the example shown above is shown in Fig. 2.2.

The oval form around player's B decision nodes indicates that the two nodes are in a single information set, that is, when player B chooses C or NC he does not know in which node he is, because he did not observe the move of player A. In this case, as players take their decisions simultaneously and are symmetric, we could simply switch the order in which the players appear in the tree. If we would remove the

		Player B	
		C	**NC**
Player A	**C**	4, 4	0, 5
	NC	5, 0	1, 1

Fig. 2.1 A game in strategic form

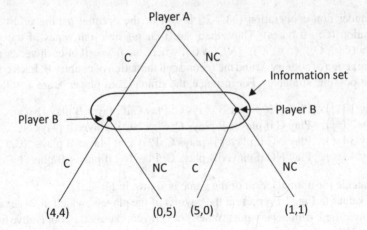

Fig. 2.2 A game in extensive form

information set, then the tree would indicate that player A would take the fishing effort decision before player B. Then, instead of a simultaneous move game, we would have a sequential game.

2.2 The Nash Equilibrium

The most widely used solution concept in game theory is due to Nash (1951) and is called the Nash equilibrium. The roots of this concept can be traced back to the works of the French mathematician Antoine Cournot in the first half of the nineteenth century (Rubinstein 1990).

> **Definition 2.4** A strategy profile $(s_1, s_2, \ldots, s_n)$ constitutes a Nash equilibrium of the game $\Gamma = [N, \{S_i\}, \{u_i(.)\}]$ if for every $i = 1, \ldots, n$, $u_i(s_i, s_{-i}) \geq u_i(s'_i, s_{-i})$, for all $s'_i \in S_i$, where s_{-i} represents the strategies of all players other than i.

In a Nash equilibrium, each player's strategy is a best response to the strategies actually played by the other players (Mas-Collel et al. 2012).

Consider Example 2.1, in which the representation in strategic form of the fishery game is shown in Fig. 2.1. In Fig. 2.3, the best responses of each player to an action of the other player are shown by underlining the payoff of his best response. Suppose player B chooses the cooperative fishing effort (C); then the best response for player A is to set the non-cooperative fishing effort (NC), which will yield him a payoff of 5. This is higher than the payoff he would get by choosing C, 4. If player B chooses NC, then again the best response of player A is to adopt NC. For any action of player A,

		Player B	
		C	NC
Player A	**C**	4, 4	0, 5̲
	NC	5̲, 0	1̲, 1̲

Fig. 2.3 The Nash equilibrium in a game in strategic form

the best response of player B is also NC. Thus, (NC, NC) forms the Nash equilibrium of the game, as no player has incentive to deviate.

There are strong arguments in favour of the Nash equilibrium as a solution concept. Among them, we can highlight the following (Mas-Collel et al. 2012):

(i) *Rational inference*: Assuming that players are rational, it is expected that they will be able to forecast the choices of the other players.

(ii) *Nash equilibrium as a necessary condition*: If players share the belief that there is an obvious (in particular, a unique) way to play a game, then it must be a Nash equilibrium.

(iii) *Nash equilibrium as a stable social convention*: If an outcome is to become a social convention, it must be a Nash equilibrium. If it were not, then individuals would start deviating from it. Thus, the Nash equilibrium can be viewed as resulting from a dynamic adjustment, which is according to the equilibrium notions in economics.

2.3 Non-cooperative Games

According to Aumann and Hart (1992), a non-cooperative game can be defined as follows.

Definition 2.5 A game is non-cooperative if commitments (agreements, promises, threats) are not enforceable.

This means that each player acts independently of the others (Maschler et al. 2013). However, it does not preclude the possibility of correlation of actions between players, exchange of information, or the existence of an observer who can make recommendations to the players about the actions to choose. The key aspect is that the correlations or recommendations are not binding. Each player has always the possibility to choose any action from the section of actions available to him.

A non-cooperative game is classified as a *static game* (or simultaneous move) when players move only once and at the same time, or a *dynamic game* when players choose actions over time. In a dynamic game, each choice of action by the players corresponds to a stage of the game and hence dynamic games are also called *multi-stage games*. Usually, stages are identified with time periods. However, a stage may not have a temporal interpretation (Fudenberg and Tirole 1991) and hence a multi-stage game can be static (see, for instance, the two-stage games presented in Chap. 6).

In multi-stage games, the concept of *subgame perfect equilibrium* is applied. In order to define it, the concept of subgame must be introduced. Let h^k denote the history of the game at stage k, that is, the actions of players before stage k (stages 1 to $k-1$). The game starting at stage k, with history h^k, is called a *subgame* and can be denoted as $\Gamma\left(h^k\right)$. Following Maschler et al. (2013), we can now define subgame perfect equilibrium.

Definition 2.6 A strategy profile $s = (s_1, s_2, \ldots, s_n)$ in an extensive-form game $\Gamma = \left[N, \{S_i\}, \{u_i(.)\}\right]$ is a **subgame perfect equilibrium** if, for every subgame $\Gamma\left(h^k\right)$, the restriction of s to the subgame is a Nash equilibrium. For every player $i \in N$, every strategy s_i, and every subgame $\Gamma\left(h^k\right): u_i\left(s|h^k\right) \geq u_i\left(s_i', s_{-i}|h^k\right)$.

Subgame perfect equilibrium is a refinement of the Nash equilibrium. It requires that the strategy is a Nash equilibrium not only of the entire game but also of all subgames. This refinement allows to reduce the number of equilibria, when the entire game has more than one Nash equilibrium. The idea behind this equilibrium concept is to rule out noncredible threats (Maschler et al. 2013). Finding the subgame perfect equilibrium of a game is undertaken through a method called *backward induction*. As the name suggests, the first step is to find the Nash equilibrium of the last stage. The second step is to compute the Nash equilibrium of the penultimate stage, given the equilibrium strategies of the last stage. The algorithm is applied successively until the Nash equilibrium of the first stage is obtained.

Another important classification is related to information. In a *game of complete information,* the players' payoff functions are common knowledge. When at least one of players does not know another player's payoff, we have a game of incomplete information (Gibbons 1992). A game is of perfect information if each player makes all decisions being perfectly informed of all the events that occurred previously (Osborne and Rubinstein 2016). If one or more players take decisions whilst not knowing the history of the game, the game is of imperfect information.

The game represented in Fig. 2.3 is a static game, in which both player A and player B choose their actions simultaneously. This is a one-stage game often referred as one-shot game. It is also a game of complete information as each player knows the payoff of the other under each possible outcome. A static game of complete information is the simplest setting used to analyse the strategic interaction of decision-makers in fisheries. However, it is usually powerful enough to provide important lessons.

Figure 2.3 represents a game which has the features of one of the most famous games: The *Prisoner's Dilemma* (see Box 2.1). For each player, the non-cooperative fishing effort (NC) is the dominant strategy, that is, independently of what the other player chooses the best response is always NC. Hence, in the equilibrium, both countries adopt a non-cooperative fishing effort. The dilemma is that both players could get a higher payoff if they would communicate and agreed to cooperate. The strategic interaction between players leads to an equilibrium which is not desirable.

Box 2.1 The *Prisoner's Dilemma*

Two individuals are arrested and accused of having committed a crime. They are kept in different cells so that they cannot communicate. Each individual has two possible decisions: Confess or Don't Confess. The prosecutor informs each of them, separately, that if both confess they will go to prison for 5 years. If only one confesses, then he will be freed for collaborating with justice and the other will be sentenced to six years in prison. If both do not confess, their sentence will be one year in prison for minor offense. The payoff of each player is given by the number of years of freedom that he will enjoy over the next six years. The strategic representation of this game is shown in Fig. 2.4.

		Individual 2	
		Don't Confess	**Confess**
Individual 1	**Don't Confess**	5, 5	0, 6
	Confess	6, 0	1, 1

Fig. 2.4 The Prisioner's Dilemma game in strategic form

The solution of this game, the Nash equilibrium, is (Confess, Confess) an outcome in which the players do not cooperate. However, the players could have reached a higher payoff if both cooperated by choosing Don't Confess.
Source: Adapted from Myerson (2004).

2.4 Cooperative Games

Following Aumann and Hart (1992), a cooperative game can be defined as follows.

Definition 2.7 A game is cooperative if commitments (agreements, promises, threats) are fully binding and enforceable.

If a cooperative game, in contrast to a non-cooperative game, is one in which the commitments are fully binding and enforceable, then this immediately raises the question of what conditions must prevail for such a binding outcome to be achieved. A subsidiary question is how the economic benefits arising from cooperation are to be shared, if cooperation is in fact realized.

In cooperative games, unlike non-cooperative games, the number of players matters. There is a great deal of difference between $n = 2$ and $n > 2$ cooperative games, with the latter being much, much more complex than the former. Chapter 4 is focussed entirely on the relatively simple $n = 2$ cooperative games. While restrictive, $n = 2$ cooperative games have value in analysing the economic management of transboundary fishery resources.

The question of the conditions that must be achieved for a $n = 2$ cooperative game to be binding, to be stable through time, reveals that cooperative games are not independent of non-cooperative games. The first condition that must be met is that each player at each point in time must be assured a payoff from cooperation at least as great as it would enjoy under non-cooperation. The payoff to each player under non-cooperation is assumed to be the payoff arising from the solution to a non-cooperative game. The second key condition is that the solution to the cooperative game must be Pareto optimal, which means that it is not possible to make one player better off without harming the other.

The optimal division of the benefits from cooperation between the two players is straightforward. The Cooperative Surplus is defined in Chap. 4 as being equal to the sum of the payoffs arising from the solution to the cooperative game minus the sum of the payoffs arising from the solution to the non-cooperative game. The optimal division is such that each player receives their non-cooperative game payoff plus one-half of the Cooperative Surplus.

When one turns to $n > 2$ cooperative games, the question of coalitions becomes paramount. Indeed, we now refer to coalitional games. If the set of players is represented by $N = \{1, 2, \ldots, n\}$, there are then 2^n possible coalitions, including the Grand Coalition (formed by all players), sub-coalitions, the players as singletons, and the empty coalition $\emptyset$. In an $n = 2$ cooperative game, there are thus four possible coalitions: the two singletons, the empty coalition and the Grand Coalition. The only coalition of real interest is the Grand Coalition. Will the two players cooperate, or will they not?

Suppose, for the sake of argument, that the cooperative game has $n = 5$ players, not an exceptionally large game. There are then 32 possible coalitions. Along with the Grand Coalition, the singleton coalitions and the empty coalition, there are 25 possible sub-coalitions. The situation is now truly complex. Maschler et al. (2013) maintain that, with $n > 2$ cooperative games, there are two key questions to be addressed:

(i)   which coalitions will succeed in forming binding agreements?
(ii) if a coalition is established, how will the benefits from cooperation be divided between/among the coalition members?

Question (i) is concerned with achieving stability within coalitions, the Grand Coalition, in particular. Question (ii) is about sharing, about achieving a "fair" division of the economic benefits from cooperation. Question (ii) is linked to (i) in the sense that manifestly "unfair" divisions will serve to undermine the stability of the coalitions.

These questions in the context of $n > 2$ cooperative games are complex and difficult. So complex are they that it is not possible to address them both in a single chapter. Chapter 5 focusses its attention on Question (ii), the sharing question.

It will, in fact, be seen from Chaps. 4 and 6, as well as Chap. 5, that the literature contains a wide set of possible sharing rules, with examples being: the Nash bargaining solution, the Shapley value, the Nucleolus and the Almost Ideal Sharing Scheme. These sharing rules rest upon different fairness concepts. Under the Nash bargaining solution, all players are seen as being equally important in achieving cooperation, with the consequence that fairness demands that they receive equal shares of the gains from cooperation. The Shapley value divides the gains according to each player's average contribution to the coalitional worth. The nucleolus maximizes the minimum gains to any possible coalition. Finally, the Almost Ideal Sharing Scheme, especially designed for coalition games, assigns to each player its payoff when defecting from the coalition, plus a share of the gains from cooperation. A detailed analysis of these concepts will be provided in Chap. 5, along with Chaps. 4 and 6.

Chapter 6 will address in detail the difficult stability question, Question (i), in detail. An issue, which has policy implications of critical importance, is the threat to the Grand Coalition created by free riding. The larger the Grand Coalition, the greater the threat.

Example 2.2 Consider a fish stock that is being exploited by two coastal states, 1 and 2. If the two coastal states succeed in cooperating, the Grand Coalition will generate an economic value of 20 monetary units. If the two do not cooperate, the payoffs from the non-cooperative game will be 6 for player 1, and 8 for player 2. Suppose that the two do cooperate; how then to divide the 20 between the two?

Let the payoffs of players 1 and 2 arising from a solution to the cooperative game be denoted as μ_1 and μ_2, respectively. It has been seen that a fundamental condition to be met, if the solution to the cooperative game is to prove to be stable, is that the payoffs from cooperation be at least as great as the payoffs from non-cooperation, now to be referred to as Threat Points. The cooperative game payoffs must also be Pareto optimal. Thus, the cooperative game payoffs must satisfy the following conditions: $\mu_1 + \mu_2 = 20$; $\mu_1 \geq 6$; $\mu_2 \geq 8$. Any pair (μ_1, μ_2) satisfying these conditions is said to belong to the Core of the game.[1]

As discussed, a possible answer to the question of how to divide the 20 between the two is to give each player its Threat Point payoff plus one-half of the Cooperative Surplus. By observation, the Cooperative Surplus in this case is equal to 20–14 = 6. Thus, each player would receive its Threat Point payoff plus 3. Hence, the players' payoffs would be $(\mu_1, \mu_2) = (9, 11)$. As will be shown in Chap. 4, the solution

---

[1]A more formal definition in the context of two-player games is provided in Chap. 4.

Fig. 2.5 A cooperative game

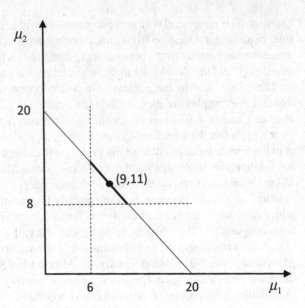

concept used in this example corresponds to the well-known sharing rule: the Nash bargaining solution.

This example is illustrated in Fig. 2.5. In the figure, the 45° negatively sloped line is referred to as the Pareto Frontier. The segment of the Frontier between the dashed lines represents the Core of the Game.

2.5 Case Study: The North Sea Herring Fishery

The North Sea herring fishery is both an ancient and important fishery, which provides us an example of both a non-cooperative fishery game and a cooperative one. Harvests in the fishery, which have been as high as 1 million tonnes per annum, have, in recent years, been in order of 500,000 t per annum (Arnason et al. 2018).

The resource was probably close to its natural equilibrium level at the end of the Second World War, due to the impact of the War upon fishing activities. Following the end of the War, the resource was fished under open-access conditions, which proved not to be a threat, until the 1960s when major advances in fishing technology led to substantial increased fishing pressure (Arnason et al. 2018).

ICES estimates that the minimum safe spawning stock biomass (SSB) of the resource is 800,000 t. In 1963, the SSB was estimated to be almost 2.2 million tonnes. The increased fishing pressure, along with a recruitment failure, saw the SSB plummet to an estimated 47,000 tonnes by 1977, with the scientists warning of possible extinction of the resource (Arnason et al. ibid.).

Consider the fishery in the late 1960s and first part of the 1970s, when ICES scientists were expressing ever intense warnings. Prior to 1977, coastal states jurisdiction extended to a maximum of 12 nautical miles. The North Sea was thus largely high seas. The North Seas herring stock was being exploited by not fewer than 14 states (Dickey-Collas et al. 2010). The fishery was under the ostensible international supervision and oversight of the Northeast Atlantic Fisheries Commission (NEAFC). NEAFC did in turn provide a basis for communication among the 14+. Nonetheless, the North Sea herring fishery game was a decidedly non-cooperative one.

NEAFC stood ready to recommend harvest reductions but could do so only if all member states agreed. There was no international inspection system, which meant that achieving a binding agreement among the 14+ players was all but impossible. Dickey-Collas et al. (2010) report that obtaining agreement among the member states on substantial catch reductions proved impossible because "...there was widespread concern that catch restrictions would not be implemented by other states" (Dickey-Collas et al. 2010, p. 1881). This was a pure Prisoner's (Fisher's) Dilemma situation, arising even though the players could effectively communicate with one another.

In Chap. 1, it was pointed out that, while the UN Convention on the Law of the Sea was not ready for signing before 1982, the fisheries issues had been settled by 1975, with the result that coastal states began implementing EEZs by 1977. On 1 January 1977, all coastal states around the North Sea implemented EEZs, with the consequence that the North Sea waters ceased at that moment to be high seas (Dickey-Collas et al. 2010). Shortly thereafter, the European Union members agreed to establish a Common Fisheries Policy (CFP) (Dickey-Collas et al. ibid.). These developments undoubtedly saved the resource.

The consequence of the implementation of the EEZs around the North Sea and the EU CFP was that the intractable 14+ player game was reduced to a two-player game, with the players being the EU and Norway. The competitive fishery game was quickly transformed into a stable two-player fishery cooperative game. With respect to the motivation of the two players to cooperate, the motivation was stark and obvious. The previous non-cooperative fishery game had threatened to drive the fishery resource to destruction.

Under the EU-Norway cooperative resource management arrangement, the resource was rebuilt, with the rebuilding of the resource, the investment in this fishery natural capital, occurring with reasonable speed. A phased-in re-opening of the transboundary fishery occurred between 1981 and 1983 (Bjørndal 1988). The fishery resource fluctuates due to natural causes (e.g. strong and weak year classes), but, since 1983, the spawning stock biomass has been safely well above the minimum safe level (Arnason et al. 2018).[2]

This cooperative fishery game, now well past its 40th anniversary, has led to a cooperative agreement, which has proven to be binding over time. That the results to date are far, far superior to the non-cooperative alternative is not to be doubted. Bjørndal and Lindroos (2004) model the non-cooperative North Sea herring fisheries game (along with the cooperative game) and conclude that the non-cooperative game,

---

[2]With the exception of a brief period in the mid-2000s (Dickey-Collas et al. 2010, p. 1882).

even with just two players, is "very destructive in the biological sense"[3] (Bjørndal and Lindroos 2004, 92). One does not need precise estimates to be able to assert, without fear of contradiction, that each player finds its cooperative payoff to be far greater than its Threat Point payoff.

References

Arnason, R., Bjørndal, T., Gordon, D., & Bezabih, M. (2018). Measuring potential rents in the North Sea herring fishery. *American Journal of Agricultural Economics, 100,* 889–905.

Aumann, R. J., & Hart, S. (1992). *Handbook of game theory with economic applications* (Vol. 1). Amsterdam: Elsevier Science Publishers.

Bjørndal, T. (1988). The optimal management of North Sea herring. *Journal of Environmental Economics and Management, 15,* 9–29.

Bjørndal, T., & Lindroos, M. (2004). International management of North Sea herring. *Environmental & Resource Economics, 29*(1), 83–96.

Dickey-Collas, M., Nash, R., Brunel, T., van Damme, C., Marshall, T., Payne., M., Corten, A., Geffen, A., Peck, M., Hatfield, E., Hinyzen, N., Enberg, K., Kell, L., & Simmonds, E. (2010). Lessons learned from stock collapse and recovery of North Sea herring: A review. *ICES Journal of Marine Science, 67*(9), 1875–1886.

Fudenberg, D., & Tirole, J. (1991). *Game theory.* Cambridge: The MIT Press.

Gibbons, R. (1992). *A primer in game theory.* Harlow: Pearson Education Limited.

Maschler, M., Solan, E., & Zamir, S. (2013). *Game theory.* Cambridge: Cambridge University Press.

Mas-Collel, A., Whinston, M., & Green, J. (2012). *Microeconomic theory.* New York: Oxford University Press.

Myerson, R. B. (2004). *Game theory: Analysis of conflict.* Cambridge, MA: Harvard University Press.

Nash, J. (1951). Non-cooperative games. *Annals of Mathematics, 54*(2), 286–295.

Osborne, M. J., & Rubinstein, A. (2016). *A course in game theory.* Cambridge, MA: Phi—MIT Press.

Rubinstein, A. (1990). *Game theory in economics.* The international library of critical writings in economics 5. Hants: Edward Elgar.

---

[3]Not just in the biological sense. If one views the resource as a form of natural capital, the implication is that non-cooperation would lead to extensive disinvestment in the natural capital that would be decidedly non-optimal. At worst, it could lead to outright liquidation of the natural capital.

Chapter 3
Introduction to Non-cooperative Fisheries Games

Abstract The purpose of the current chapter is to introduce basic non-cooperative fisheries games. In non-cooperative games, players act competitively and choose best responses to the actions of the other players. The chapter starts with a classic static game in which players use one-shot strategies. This game illustrates, among other things, the fact that a large number of players competing for a fish stock can have grave consequences for the stock level and the resource rent. Then, another classic game is presented: a dynamic game with feedback strategies. This game demonstrates that the most efficient players tend to harvest the stock down to a level where it ceases to be in the best interests of the other players to remain in the fishery. The chapter concludes with a case study on the South Tasman Rise trawl fishery, which illustrates the fact that the outcomes of non-cooperation may be severe.

A game in which commitments (agreements, promises, threats) are not enforceable is defined as non-cooperative (see Definition 2.5). This implies that each player acts independently of the others. As shown in Chap. 2, non-cooperative games can be divided into static and dynamic games. A static game is by definition timeless, and will be seen in our case to be one in which certain constraints are imposed. In a dynamic game, time is taken into account explicitly, and the aforementioned constraints are not imposed. This chapter presents two classic non-cooperative fisheries games, one static and another dynamic, which have been widely used in the fisheries economics literature.

3.1 A Static Game

This section starts by presenting a static, or steady state, game with symmetric players. The simple two-player case is first discussed, which is then generalized for n-players. The outcomes are illustrated through a numerical example. Then, the case of a static game with asymmetric players is explored, following the same steps.

3.1.1 Symmetric Game

Two-Player Game

Consider a simple non-cooperative fishery game, in which two players, 1 and 2, exploit a common fish stock. These players are identical in all aspects (fish price, harvesting costs and technology) and are thus symmetric. Each player chooses its actions in order to maximize its individual payoff, given the actions of the other players. Let us note in passing that the players can be fishers, fleets, countries or groups of countries.

This game, and all of the static games to follow, is cast within the framework of the famous Gordon–Schaefer model (Gordon 1954). That static model focuses on the revenues, costs and resource rent arising from harvesting sustainable yields from the fishery resource. Let us be reminded that the sustainable yield/ harvest associated with any given level of the biomass, $X(t)$, is equal to the net natural growth, $F(X(t))$, of the biomass at that level. As we shall see, the economic net return from the fishery to player 1 will be determined by player 1's share of the sustainable yield/harvest, whatever that might be. What holds true for player 1, holds true for player 2.[1]

From the Gordon–Schaefer model, $F(X(t))$, takes the following functional form, based upon the logistic growth function:

$$F(X(t)) = rX(t)\left(1 - \frac{X(t)}{k}\right), \tag{3.1}$$

where r is the intrinsic growth rate and k is the carrying capacity of the environment. The growth is zero when $X = 0$ or $X = k$ and achieves its maximum value, often referred to as the maximum sustainable yield (MSY), when $X = k/2$.

It is next assumed that the harvest production function of player i is based upon the Gordon–Schaefer harvest production function, and thus takes the following form:

$$H_i(t) = qE_iX(t), \tag{3.2}$$

where q is the catchability coefficient and E is the rate of fishing effort.

If harvesting is taking place on a sustained yield basis, then we have

$$qE_1X(t) + qE_2X(t) = rX(t)\left(1 - \frac{X(t)}{k}\right). \tag{3.3}$$

For any given set of Es, E_1 and E_2, we can from Eq. (3.3), determine the corresponding steady-state level of $X(t)$.[2] We have

---

[1] That is to say, the players combined are to be seen as "skimming off" the net natural growth of the resource.

[2] i.e. for any given set of Es, solve for $X(t)$.

$$X = k\left(1 - \frac{q(E_1 + E_2)}{r}\right). \tag{3.4}$$

Since the fish stock $X(t)$ is common to the players, it is clear from Eq. (3.4) that an increase of fishing effort by one player will lead to a decrease in the steady-state level of $X(t)$. Herein lies the source of strategic interaction in the game.

Before proceeding to the game, let us first digress and ask what the outcome would be, if the fishery were under the control of a sole owner, a single player. This will provide us with a benchmark, against which we can measure the effects of non-cooperation between/among two or more players.

The sole owner's goal can be seen as attempting to maximize the following:

$$\max_E \pi = pH - cE, \tag{3.5}$$

where p is the price per unit harvested and c is the unit cost of effort determined as the opportunity cost. By assumption, both p and c are constants, in keeping with the Gordon–Schaefer model (Gordon 1954). Thus, π is to be seen as resource rent.

But, harvesting is taking place on a sustained yield basis. From Eqs. (3.2) and (3.4),[3] by substituting for $X(t)$ in Eq. (3.2), we get the following:

$$H = qEk\left(1 - \frac{qE}{r}\right). \tag{3.6}$$

Hence, the sole owner's maximization problem can be expressed as

$$\max_E \pi = pqEk\left(1 - \frac{qE}{r}\right) - cE. \tag{3.7}$$

Taking the first-order condition yields the optimal rate of fishing effort, the one that will maximize the sole owner's sustainable resource rent, the sole owner's maximum economic yield (MEY).[4] We have

$$E = \frac{r}{2q}(1 - b), \tag{3.8}$$

where parameter $b = \frac{c}{pqk}$ is commonly referred to as an inverse efficiency parameter (Mesterton-Gibbons 1993), because it increases with the cost per unit of effort and decreases with the price and the catchability coefficient.

Let us now refer to the sole owner's optimal E from Eq. (3.8) as E^{opt}.

From Eq. (3.4),[5] we can determine the optimal level of $X(t)$, X^{opt}, in that X^{opt} is associated with the maximum sustainable resource rent. We have

---

[3]With appropriate adjustment.
[4]See Bjørndal and Munro (2012), Chap. 2.
[5]Once again, with all necessary adjustments being made.

$$X^{\text{opt}} = k\left(1 - \frac{q E^{\text{opt}}}{r}\right). \tag{3.9}$$

With the digression complete, we can proceed to the non-cooperative game. The sole owner is joined by a second player, with the result that we now have players 1 and 2. Let us not forget that, by assumption, the players are symmetric.

In the game to follow, it is assumed that all players choose their fishing effort levels simultaneously. These games are often called one-shot games, as the fishing effort decision is taken only once and simultaneously by all players. If we return to Eqs. (3.3) and (3.4), it is clear that the combined effort levels lead to a steady-state stock level.

We now have for player i:

$$H_i = q E_i k\left(1 - \frac{q(E_i + E_j)}{r}\right), \quad \text{where} \quad i \neq j \in \{1, 2\}. \tag{3.10}$$

Hence, the maximization problem of player i is to be written as

$$\max_{E_i} \pi_i = pq E_i k\left(1 - \frac{q(E_i + E_j)}{r}\right) - c E_i. \tag{3.11}$$

Thus, π_i is now to be interpreted as player i's share of the sustainable resource rent. Next, taking the first-order condition yields the following optimal fishing effort for an individual player, given the fishing effort of the other players:

$$E_i = \frac{r}{2q}(1 - b) - \frac{E_j}{2}, \quad i \neq j \in \{1, 2\}, \tag{3.12}$$

where the strategic interaction becomes obvious.

These optimal solutions can be seen as reaction functions or best-response functions:

$$E_1 = \frac{r}{2q}(1 - b) - \frac{E_2}{2}, \tag{3.13}$$

$$E_2 = \frac{r}{2q}(1 - b) - \frac{E_1}{2}. \tag{3.14}$$

That is to say, Eq. (3.13) indicates the optimal E_1, given E_2; Eq. (3.14) indicates the optimal E_2, given E_1.

From Eqs. (3.13) and (3.14), we conclude that the reaction functions are linear in the other player's effort. This allows us to analyse two extreme solutions. A player chooses the sole owner optimum fishing effort, leading to the maximization of sustainable resource rent, if the other player chooses zero effort (see Eq. 3.8). The other extreme is when the other player chooses the open-access fishing effort level, leading to the complete dissipation of sustainable resource rent. It is then optimal for the player to choose zero effort.

The analytical solution for the equilibrium fishing efforts is obtained by inserting player 2's reaction function into player 1's reaction function, that is, solving a system of two simultaneous equations. In this case, as players are symmetric, their effort level will be identical:

$$E_1 = \frac{r}{3q}(1 - b) = E_2. \tag{3.15}$$

Summing the fishing effort of the two players yields the aggregate fishing effort:

$$AE = \frac{2r}{3q}(1 - b). \tag{3.16}$$

The corresponding steady-state stock level, which we shall denote as X^{AE}, is given by

$$X^{AE} = k\left(1 - \frac{qAE}{r}\right). \tag{3.17}$$

The individual payoff, π_i, can be determined through Eq. (3.11).

Now compare Eq. (3.16) with Eq. (3.8). Upon so doing, it becomes immediately apparent that $AE > E^{opt}$.[6] If $E > E^{opt}$, then it becomes equally apparent (see Eq. 3.17) that $X^{AE} < X^{opt}$. The sustainable resource rent will, of course, be below the maximum, the MEY level. We can thus set forth the consequences of there being in the fishery two players, rather than one, with the two playing a non-cooperative game. The consequences are: (i) excessive fishing effort, (ii) resource depletion and (iii) partial dissipation of resource rent.

We now turn to $n > 2$ player non-cooperative games, where it is guaranteed that matters will get worse.

n-Player Game

We now generalize the game for any number of players, $n \geq 2$. Denote the fishing effort of player i by E_i, where $i \in \{1, 2, \ldots, n\}$, then, similarly to Eq. (3.4), for the case of two players, the steady-state stock depends on the sum of the fishing efforts over all players:

$$X = k\left(1 - \frac{q \sum_{i=1}^{n} E_i}{r}\right). \tag{3.18}$$

The maximization problem of player i can thus be written as

---

[6]Simply note that $\frac{2}{3} > \frac{1}{2}$.

$$\max_{E_i} \pi_i = pq E_i k \left(1 - \frac{q\left(E_i + \sum_{j \neq i} E_j\right)}{r} \right) - c E_i. \tag{3.19}$$

Taking the first-order condition yields the reaction function of player i:

$$E_i = \frac{r}{2q}(1 - b) - \frac{\sum_{j \neq i} E_j}{2}, \quad i \in \{1, 2, \ldots, n\}. \tag{3.20}$$

The optimum fishing efforts are obtained by solving the n simultaneous equations in Eq. (3.20):

$$E_i = \frac{r}{(n+1)q}(1 - b). \tag{3.21}$$

Thus, the aggregate fishing effort is given by

$$AE = \sum_{i=1}^{n} E_i = \frac{nr}{(n+1)q}(1 - b). \tag{3.22}$$

The aggregate fishing effort increases with the number of players, as $\frac{\partial AE}{\partial n} > 0$. Moreover, as the number of players, n, approaches infinity, the aggregate fishing effort converges to $AE = \frac{r}{q}(1 - b)$, which is the aggregate fishing effort under open access, or to use H. Scott Gordon's term, Bionomic Equilibrium (Gordon 1954).

To demonstrate this, let us be reminded that Bionomic Equilibrium is achieved when two conditions are met. The first, in keeping with our discussion so far, is that the resource is being harvested on a sustained yield basis. The second is that the resource rent from the fishery is fully dissipated. We can express these conditions as follows:

$$\pi = pH - c AE^{OA} = 0, \tag{3.23}$$

where AE^{OA} denotes aggregate fishing effort under open access, Bionomic Equilibrium, and where H is seen to be sustainable harvest, $H = q AE^{OA} X^{OA}$. Drawing upon Eq. (3.18), we can re-express Eq. (3.23) as

$$\pi = pq AE^{OA} k \left(1 - \frac{q AE^{OA}}{r} \right) - c AE^{OA} = 0. \tag{3.23a}$$

Equation (3.23a) can be solved for AE^{OA}.[7] Upon so doing, we have

$$AE^{OA} = \frac{r}{q}\left(1 - \frac{c}{pqk} \right), \tag{3.24}$$

---

[7] Obtaining the solution can be safely left to the reader.

which can be re-expressed as

$$AE^{OA} = \frac{r}{q}(1 - b).$$ (3.24a)

Now return to Eq. (3.18). Replacing $\sum_{i=1}^{n} E_i$ with AE^{OA}, and then substituting for AE^{OA} from Eq. (3.24), we can solve for the Bionomic Equilibrium steady-state biomass, to be denoted as X^{OA}. We have

$$X^{OA} = \frac{c}{pq}.$$ (3.25)

If we compare the aggregate effort level from a n-player non-cooperative game in Eq. (3.22) with the aggregate effort level under open access, Bionomic Equilibrium, (Eq. 3.24a), we conclude that, as n approaches infinity, the aggregate fishing effort, AE, does indeed converge to the open access, Bionomic Equilibrium level—Q.E.D.

Now return to Eq. (3.22). The corresponding steady-state biomass can be determined by substituting $\sum_{i=1}^{n} E_i$ in Eq. (3.18) by AE from Eq. (3.22).We then have[8]:

$$X = k\left(1 - \frac{n}{n+1}(1 - b)\right).$$ (3.26)

By taking partial derivatives we can conclude that, as expected, the equilibrium stock level increases with the carrying capacity of the environment, k. It also increases with the inverse efficiency parameter, $b = \frac{c}{pqk}$, which means that the lower the efficiency of the fishery the higher the equilibrium stock level. High costs, low prices or low catchabilities lead to stock conservation. Moreover, we can also conclude that the equilibrium stock level decreases with the number of players.

To have a complete picture of the game, we need to compute the equilibrium payoffs. For that, we insert the equilibrium fishing efforts (Eq. 3.21) into the payoff function in Eq. (3.19). This yields the payoff of player:

$$\pi_i = \frac{rpk}{(n + 1)^2}(1 - b)^2.$$ (3.27)

Finally, the aggregate payoff of the n players is

$$A\pi = \frac{nrpk}{(n + 1)^2}(1 - b)^2.$$ (3.28)

We can conclude that as the number of players increases both individual and aggregate payoffs decrease ($\frac{\partial \pi_i}{\partial n} < 0$ and $\frac{\partial A\pi}{\partial n} < 0$).

---

[8]The reader can verify that, as n approaches infinity, the steady-state biomass, X, approaches X^{OA}, as given by Eq. (3.25).

Table 3.1 Outcomes of static games with symmetric players

	Number of player (n)					
	1	2	3	10	100	∞
E_i	93.75	62.50	46.88	17.05	1.86	$\approx 0^a$
AE	93.75	125.00	140.63	170.45	185.64	187.50
X	62.50	50.00	43.75	31.82	25.74	25.00
π_i	42.19	18.75	10.55	1.39	0.02	0.00
$A\pi$	42.19	37.50	31.64	13.95	1.65	0.00

[a]The fishing effort of each player will be infinitesimally small

Earlier, we set forth the consequences of there being two players, rather than one, in the fishery, with the two playing competitively. Our overall conclusion is that these consequences—excessive fishing effort, resource depletion and resource rent dissipation—are steadily intensified as the number of players playing competitively increases. In the limit, as that number increases, the fishing effort will reach a point where, in this model, $= 2E^{opt}$,[9] with the consequence that resource rents, at both the individual and the aggregate levels, will be fully and completely dissipated.

Example 3.1 Symmetric players. To illustrate the static non-cooperative game model with symmetric players, we take a specific numerical example. Assume that the following parameter values hold: $r = 1$; $k = 100$; $q = 0.004$; $p = 3$ and $c = 0.3$ Then, drawing upon the formulas derived so far, we can derive the outcomes of the game for different number of symmetric players as shown in Table 3.1.

This example clearly shows that, as the number of players increases, the individual fishing effort levels decrease, but the aggregate level of fishing increases. As a consequence, the equilibrium stock level is reduced. Furthermore, both individual and aggregate payoffs decrease. Hence, the larger is the number of players in this non-cooperative game, the worse are the outcomes in both biological and economic terms.

3.1.2 Asymmetric Game

Let us now consider that players are different and hence asymmetric. We introduce asymmetry in terms of the unit cost of fishing effort.[10]

---

[9]Compare Eq. (3.24a) with Eq. (3.8).

[10]Asymmetry could also be introduced on two other model parameters: p (price) and q (catchability). Another potential source of asymmetry could arise from different information/perceptions about the stock size and/or biological parameters.

Two-Player Game

Again let us start out with a two-player game. Assume that the players have different unit costs of efforts, c_i, but are identical with respect to all the other parameters. Then the maximization problem of each player becomes

$$\max_{E_i} \pi_i = pH_i - c_i E_i. \tag{3.29}$$

We follow the same procedure as in the symmetric case. By inserting the steady-state stock from Eq. (3.4) to the harvest function in Eq. (3.2), and taking the first-order conditions, we obtain the following reaction functions:

$$E_1 = \frac{r}{2q}(1 - b_1) - \frac{E_2}{2} \tag{3.30}$$

$$E_2 = \frac{r}{2q}(1 - b_2) - \frac{E_1}{2}. \tag{3.31}$$

Solving the system of Eqs. (3.30) and (3.31) gives the Nash equilibrium:

$$E_1 = \frac{2r}{3q}(1 - b_1) - \frac{r}{3q}(1 - b_2). \tag{3.32}$$

$$E_2 = \frac{2r}{3q}(1 - b_2) - \frac{r}{3q}(1 - b_1). \tag{3.33}$$

Summing the fishing effort of the two players yields the aggregate fishing effort:

$$AE = \frac{r}{3q}[(1 - b_1) + (1 - b_2)]. \tag{3.34}$$

In this case, the fishing efforts of the players are no longer the same. The more efficient is a player (lower parameter b), in relation to the other player, the higher is its equilibrium fishing effort. The converse is, of course, equally true.

Note that the above equations only hold for interior solutions. The introduction of asymmetry complicates the analysis as one needs to determine whether both countries find it optimal to be active in the fishery. If the relative efficiency difference between the players is great enough, then the less efficient player will find it optimal not to enter the fishery. We would then, of course, end up with a sole owner optimum.

n-Player Game

The model is now extended to any number of players, $n \geq 2$. As in the n-player symmetric game, the steady-state fish stock is given by

$$X = k\left(1 - \frac{q\sum_{i=1}^{n} E_i}{r}\right). \tag{3.35}$$

The maximization problem of player i can thus be written as

$$\max_{E_i} \pi_i = pqE_ik\left(1 - \frac{q\left(E_i + \sum_{j \neq i} E_j\right)}{r}\right) - c_iE_i. \tag{3.36}$$

The reaction function of each player i is given by first-order condition of this maximization problem:

$$E_i = \frac{r}{2q}(1 - b_i) - \frac{\sum_{j \neq i} E_j}{2}, \quad i \in \{1, 2, \ldots, n\}, \tag{3.37}$$

where $b_j = \frac{c_j}{pqk}$.

Solving simultaneously the n equations in Eq. (3.32) gives the following equilibrium fishing efforts, assuming interior solutions:

$$E_i = \frac{nr}{(n+1)q}(1 - b_i) - \sum_{j \neq i} \frac{r}{(n+1)q}(1 - b_j). \tag{3.38}$$

It can thus be concluded that the equilibrium fishing effort of a given player depends negatively on its unit cost of effort, c_i, and positively on the costs of the remainder players.

The aggregate fishing effort over all players is obtained by summing Eq. (3.38) over all players, $i = 1, 2, \ldots, n$.

$$AE = \sum_{i=1}^{n} E_i = \frac{r}{(n+1)q} \sum_{i=1}^{n} (1 - b_i). \tag{3.39}$$

As expected, the aggregate fishing effort depends negatively on unit costs of effort of the different players. Note that regarding the inverse efficiency parameters, the aggregate fishing effort only depends on the sum of these parameters, $\sum_{i=1}^{n} b_i$, and consequently on its average value. The level of dispersion, or asymmetry, of these parameters does not affect the aggregate fishing effort, as long as the average value is constant.

Inserting the aggregate fishing effort (Eq. 3.39) into Eq. (3.35) yields the equilibrium stock level:

$$X = k\left(1 - \frac{1}{n+1} \sum_{i=1}^{n} (1 - b_i)\right). \tag{3.40}$$

By taking partial derivatives, we can conclude that the equilibrium stock level increases with the carrying capacity of the environment, k, and with the inverse efficiency parameters, $b_i = \frac{c_i}{pqk}$. As with the aggregate fishing effort, the stock level is affected by the average value of parameter b_i but not by its asymmetry.

Finally, let us compute the equilibrium payoffs. For that, we insert the equilibrium fishing efforts (Eq. 3.38) into the payoff function in Eq. (3.36). This yields the payoff of player i:

$$\pi_i = \frac{rpk}{(n+1)^2}\left(1 - nb_i + \sum_{j\neq i} b_j\right)^2. \tag{3.41}$$

This payoff depends negatively on the number of players n. More competitors means a lower payoff for each player. Regarding the effect of the efficiency level, measured by parameter b, it can be concluded that the less efficient is a given player i, that is the higher b_i, the lower is its payoff. The converse is also true in that the less efficient are its competitors, the higher is the payoff to player i.

The aggregate payoff of the players is obtained by summing the individual payoffs:

$$A\pi = \frac{rpk}{(n+1)^2}\sum_{i=1}^{n}\left(1 - nb_i + \sum_{j\neq i} b_j\right)^2. \tag{3.42}$$

Example 3.2 Asymmetric players. To illustrate the game with asymmetric players, we again use a numerical example. Let $r = 1$, $k = 100$, $q = 0.004$ and $p = 3$, as in Sect. 3.1.1. Regarding the unit costs of effort, we consider a benchmark case of symmetry, $c_1 = c_2 = 0.3$, and introduce cost asymmetry whilst keeping the average value constant, $\frac{c_1+c_2}{2} = 0.3$. The outcomes of the game are shown in Table 3.2.

This example shows that as cost asymmetry increases, the fishing effort of player 1, the one with the lowest cost, increases whereas the opposite occurs to the fishing effort of player 2. It turns out that the increase in the fishing effort of player 1 is exactly offset by the decrease in the fishing effort of player 2. Hence, the aggregate fishing effort does not vary. In fact, as it be concluded from Eq. (3.39), the aggregate

Table 3.2 Outcomes of static games with asymmetric players

	Costs per unit of effort			
	$c_1 = c_2 = 0.3$	$c_1 = 0.25, c_2 = 0.35$	$c_1 = 0.2, c_2 = 0.4$	$c_1 = 0.15, c_2 = 0.45$
E_1	62.50	72.92	83.33	93.75
E_2	62.50	52.08	41.67	31.25
AE	125.00	125.00	125.00	125.00
X	50.00	50.00	50.00	50.00
π_1	18.75	25.52	33.33	42.19
π_2	18.75	13.02	8.33	4.69
$A\pi$	37.50	38.54	41.67	46.88

fishing effort does not depend on the asymmetry of the costs per unit of effort, as long as their average value is constant. Since cost asymmetry does not affect the aggregate fishing cost it also does not impact on the stock level.

What can we say regarding payoffs? Asymmetry increases the payoff of low-cost player and decreases the payoff of the high-cost player. Moreover, it increases the aggregate resource rent, as a larger share of fishing effort is undertaken by the low-cost player.

In conclusion, it can be said that the static non-cooperative fishery game provides us with an analytically tractable model, in that it is relatively easy to calculate the relevant Es, Xs and payoffs. As will be seen in Chaps. 5 and 6, this tractable model has been extensively applied in the analysis of coalition formation. Furthermore, the model has considerable policy relevance to policymakers, such as demonstrating the effect of increasing levels of fishing effort upon long-run stocks.

That said, the static non-cooperative game does not explicitly take into account time. Arising from this in part, the model is subject to a constraint in the sense that the economic returns to the players are based solely upon sustainable harvests. With this in mind, we turn now to a discussion of dynamic non-cooperative fishery games, in which, by definition, time is explicitly taken into account, and in which the aforementioned constraint is fully relaxed.

3.2 Dynamic Non-cooperative Fisheries Game

In turning from static non-cooperative fishery games to dynamic ones, we shall focus on the dynamic non-cooperative game introduced by Clark (1980).

In setting the stage for the discussion of the static non-cooperative fishery games, we considered first the case in which the fish stock is harvested by a single player, a sole owner, thereby providing us with a benchmark. We shall do exactly the same for our discussion of dynamic non-cooperative fishery games.

Then, as with our discussion of static non-cooperative fishery games, we introduce a second player and thus have a two-player game. Two sub-cases will be explored, with the first being one in which both players are symmetric, and the second being one in which the players are asymmetric. Finally, we generalize to cases in which $n \geq 2$.

3.2.1 The Sole Owner Optimum

The discussion of the sole owner optimum in the case of static non-cooperative games was, it will be recalled, based upon the famous Gordon–Schaefer model (Gordon 1954). The discussion of the sole owner optimum in the case of dynamic non-cooperative games will be based upon a model that is essentially a dynamic version of the Gordon–Schaefer model.

To get us started, consider the following situation. Suppose that, somehow, some way, we have determined the optimal level of $X(t)$, but that the current level of $X(t)$ is below that level. How do we get from the current level of $X(t)$ to the optimal level? In the static economic model of the fishery that question is glossed over. In the dynamic economic model, where time is taken into account explicitly, that question must be addressed, it cannot be ignored. Furthermore, with time being taken into account, recognition must be given to the fact that the sole owner may discount future returns from the fishery. It will be seen that the sole owner will have good reason to do just that.

In terms of the economics, we now view $X(t)$ explicitly as a form of natural capital. The question as to the optimal level of $X(t)$ is thus seen as a theory of capital question. There is an accompanying theory of investment question. If the current level of $X(t)$ is below/above the optimal level, how rapidly do we want to invest/disinvest in $X(t)$, as we approach the optimal level?

The dynamic model of the fishery for the single player (e.g. a fleet, a country) that we shall consider is the simplest of such models. It is a linear autonomous model; autonomous in the sense that the parameters are assumed to be invariant over time. It is also assumed that the labour and produced capital involved in the fishery are both perfectly malleable with respect to the fishery.[11]

The dynamics of the stock through time is set forth as follows:

$$\frac{dX}{dt} = F(X(t)) - H(t). \tag{3.43}$$

This basically tell us that the variation of the stock through time is equal to its growth net of the harvest.

The harvest production function is the same as the one presented for the static game[12]:

$$H(t) = qE(t)X(t). \tag{3.44}$$

The reader will recall that q is the catchability coefficient, and that E is the rate of fishing effort. In contrast to the static model, there is no assumption that harvesting is taking place on a sustainable yield basis. All harvests through time count, be they sustainable or unsustainable.

We now commence with two preliminary conditions:

$$X(0) = X_0, \quad X(t) \geq 0. \tag{3.45}$$

---

[11] This means that it is assumed that the labour and produced capital can be easily and costlessly be moved in and out of the fishery. This is a strong assumption.

The model to be presented owes its origins to Clark and Munro (1975), which thus appeared almost a quarter of a century after the famous Gordon (1954) article. In addition to Clark and Munro (1975), see any of the editions of *Mathematical Bioeconomics* (Clark 1976, 1990, 2010), and Bjørndal and Munro (2012) for further discussion of the model and its more complex versions.

[12] See Eq. (3.2).

These simply state that we must have a starting point, $X(0)$, and that we cannot have negative stocks.

Now, the net economic return from the fishery at each point in time is given by

$$\pi(X(t), E(t)) = pH(t) - cE(t) \tag{3.46}$$

where π stands for net economic return, p is the fish price and c is the unit cost of effort. As before, it is assumed that both p and c are constant, and are, moreover, constant through time. Thus, as before, the net economic return is to be seen as resource rent.

The objective of the sole owner is to adopt a harvest programme through time that will maximize the present value of the net economics returns, resource rent, in our specific case, through time. Technically, we are presented with a linear optimal control problem, where $X(t)$ is the "state" variable, the variable to be controlled through time, with economists viewing the "state" variable as a form of natural capital. The first question to be settled is the optimal size of $X(t)$ through time—a capital theory question.

As well as there being a "state" variable, there must also be a "control" variable, enabling the sole owner to control the evolution of $X(t)$ through time. We have a choice of either $H(t)$ or $E(t)$. In light of our discussion of the static models in the previous section, we find it most convenient to let $E(t)$ be the "control" variable.

With that in mind, we re-express Eq. (3.46) as

$$\pi(X(t), E(t)) \doteq (pqX(t) - c)E(t). \tag{3.47}$$

Our objective functional can then be written as

$$PV(E(t)) = \int_{0}^{+\infty} e^{-\delta t}(pqX(t) - c)E(t)dt, \tag{3.48}$$

where PV stands for present value and δ is the discount rate of the sole owner.

The mathematicians require that there is an upper limit be set to $E(t)$, so to oblige, we have

$$0 \le E(t) \le E^{\max}, \tag{3.49}$$

where $E^{\max}$ represents the maximum effort capacity.

Our problem can now be expressed as maximizing $PV(E(t))$, subject to various constraints. We have

$$\max_{E(t)} PV(E(t)) = \int_{0}^{+\infty} e^{-\delta t}(pqX(t) - c)E(t)dt$$

$$\text{s.t.} \quad \frac{dX}{dt} = F(X(t)) - qE(t)X(t) \tag{3.50}$$

$$X(0) = X_0, X(t) \geq 0$$

$$0 \leq E(t) \leq E^{\text{max}}.$$

where s.t. are the initials of "subject to".

We shall denote the optimal level of $X(t)$ in our dynamic model as X^*—the answer to our theory of capital question. It can be shown that once X^* is achieved it will be a steady-state level.[13] Once X^* is achieved, the sole owner will harvest thereafter on a sustainable yield basis.

It can be shown that X^* is given to us by the solution of the following equation[14]:

$$F'\left(X^*\right) - \frac{c'(X^*)F(X^*)}{p - c(X^*)} = \delta, \tag{3.51}$$

where $c(X^*) = \frac{c}{qX}$.

The LHS of Eq. (3.51) is the internal rate of return, or the own rate of interest of the resource, being the rate of return on the marginal investment in the natural capital in the form of X. This rate of return has two components, $F'(X^*)$, the impact of the investment on sustainable harvests and $-\frac{c'(X^*)F(X^*)}{p-c(X^*)}$ the impact of the investment upon harvesting costs. Note that $c(X) = \frac{c}{qX} = \frac{cE}{qEX}$ denotes the unit harvesting cost, and $c'(X) < 0$.

Equation (3.51) essentially states invest in the natural capital $X(t)$ up to the point that the rate of return on the marginal investment in this form of capital is equal to δ, where δ is best thought of as the rate of return on alternative investments of equal risk open to the sole owner, and is thus an opportunity cost (Clark and Munro 2017).

Corresponding to X^*, there is an optimal level of $E(t)$, which we shall denote as E^*. Once X^* is achieved, we have[15]:

$$E^* = \frac{F(X^*)}{qX^*}. \tag{3.52}$$

We next have what the mathematicians call the optimal approach path question, or again what economists would call the theory of investment question. To repeat, if $X(t) \neq X^*$, how rapidly should we invest or disinvest in $X(t)$? The most rapid rate of investment would be given by setting $E(t)$ to $E(t) = 0$, while the most rapid rate of disinvestment would be given by setting $E(t)$ to $E(t) = E^{\text{max}}$. The answer to our theory of investment question is given to us by the following:

---

[13]This is due to the autonomous nature of our optimal control problem.
[14]See no. 11.
[15]This comes from the fact that at $X(t) = X^*$ harvesting takes place on a sustained yield basis, i.e. $H^* = F(X^*) = qE^*X^*$.

$$E(t) = \begin{cases} 0 & \text{if } X(t) < X^* \\ \frac{F(X^*)}{qX^*} & \text{if } X(t) = X^* \\ E^{\max} & \text{if } X(t) > X^* \end{cases} . \tag{3.53}$$

Equation (3.53) states that, if $X(t) < X^*$, it is optimal to invest in $X(t)$ at the maximum rate. If $X(t) > X^*$, it is optimal to disinvest in $X(t)$ at the maximum rate. If $X(t) = X^*$, the optimal rate of investment/disinvestment in $X(t)$ is equal to zero (no surprise).[16]

Finally, it can be shown that, if the Schaefer model holds, it is possible to calculate X^* from Eq. (3.51). This is given by the following equation (Clark 1990, p. 45):

$$X^* = \frac{k}{4}\left[\left(\frac{c}{pqk} + 1 - \frac{\delta}{r} \right) + \left\{ \left(\frac{c}{pqk} + 1 - \frac{\delta}{r} \right)^2 + \frac{8c\delta}{pqkr} \right\}^{1/2} \right]. \tag{3.54}$$

This is a formidable equation, to say the least.[17]

Next, let us comment on X^* and how it relates to X^{opt}, arising from the static model, and then comment on X^{OA}, as seen from the dynamic perspective. First with regards to X^*, we can have $X^* = X^{\text{opt}}$, but if and only if, $\delta = 0$.[18] If $\delta > 0$, the normal case, then $X^* < X^{\text{opt}}$.

The X^{OA} in the dynamic case is exactly the same as it is in the static case, the Bionomic Equilibrium, with resource rent fully dissipated. In the dynamic case, we can have $X^* = X^{OA}$, but, if and only if, $\delta = \infty$. We interpret $\delta = \infty$ to mean that the player finds itself in a situation in which it has no assurance of enjoying a net economic return from the fishery beyond today. This is about to become relevant in the non-cooperative game to follow.

3.2.2 Symmetric Game

Two-Player Game
Now let us suppose that the hitherto sole owner is joined by a second player, and that the resultant two-player game is non-cooperative. For simplicity, we assume for now that the two players, denoted by the index $i = 1, 2$, are symmetric. In this context,

---

[16]The optimality of this extreme investment/disinvestment policy rests upon the linearity assumptions and the assumption that the labour and produced capital employed in the fishery is perfectly malleable. If either the linearity assumptions or the perfect malleability assumption do/does not hold, the extreme investment/disinvestment policy is decidedly non-optimal. See Clark and Munro (1975); Clark et al. (1979); Clark and Munro (2017).

[17]The attraction of the static models becomes manifest.

[18]If $\delta = 0$, the costs of investing in $X(t)$ and temporary gains from disinvesting in the resource come to be seen as trivial. That said, making the case for setting δ equal to zero is an exceedingly difficult undertaking for the sole owner. For a player in a non-cooperative fishery game, the undertaking is impossible (Clark and Munro 2017).

the dynamics of the stock can be expressed as follows:

$$\frac{dX}{dt} = F(X(t)) - H_1(t) - H_2(t),$$ (3.55)

where H_i stands for the harvest of player i.

In this setting, each player $i = 1, 2$ maximizes the present value of its share of the net economic returns from the fishery, given the fishing effort of the other player through time, subject to the stock dynamics, initial stock size and fishing effort capacity:

$$\max_{E_i(t)} PV_i\big(E_i(t), E_j(t)\big) = \int_{0}^{+\infty} e^{-\delta t}(pqX(t) - c)E_i(t)dt$$

$$\text{s.t.} \quad \frac{dX}{dt} = F(X(t)) - qE_i(t)X(t) - qE_j(t)X(t)$$ (3.56)

$$X(0) = X_0, X(t) \geq 0$$

$$0 \leq E_i(t) \leq E^{\max},$$

where j represents the other player.

Next let us observe that each player, 1 and 2, will have its own perception of what the optimal level of $X(t)$ would be, if it were sole owner, and of the Bionomic Equilibrium stock level. Thus, we have $X_1^*; X_1^{OA}$; and $X_2^*; X_2^{OA}$.

As we assume that the players are symmetric, all parameters are identical for the two players. The consequence is that $X_1^* = X_2^*; X_1^{OA} = X_2^{OA}$.

Solving this problem for both players yields the Feedback Nash equilibrium fishing effort strategies which correspond to a most rapid approach to the open access, Bionomic Equilibrium, stock level, $X^{OA} = \frac{c}{pq}$, in which the resource rent from the fishery is fully dissipated:

$$E_i(t) = \begin{cases} 0 & \text{if } X(t) < X^{OA} \\ \alpha_i \frac{F(X^{OA})}{qX^{OA}} & \text{if } X(t) = X^{OA} , \\ E^{\max} & \text{if } X(t) > X^{OA} \end{cases}$$ (3.57)

where $0 \leq \alpha_i \leq 1$, and $\sum_{i=1}^{2} \alpha_i = 1$.

The equilibrium effort strategies are called Feedback, as they depend on the stock level. If the stock level is below the open-access stock level, X^{OA}, then players cease fishing to avoid losses. On the other hand, if the stock is above X^{OA} then players adopt their maximum fishing effort as resource rents are being garnered. When the stock is at the Bionomic Equilibrium, $X(t) = X^{OA}$, the resource rent of each player is nil, $(pqX^{OA} - c)E(t) = 0$ and so is the marginal resource rent from an additional unit of fishing effort, $pqX^{OA} - c = 0$. Thus, each player is indifferent in adopting any effort level in the range $[0, E^{\max}]$. Consequently, the Nash equilibrium of this game is not unique, as any division of the aggregate effort level, $\frac{F(X^{OA})}{qX^{OA}}$, among the players is an equilibrium.

To sum up, in the equilibrium strategies of the two-player symmetric game, there is a most rapid approach to the Bionomic Equilibrium stock level, in which the resource rent is dissipated. Regarding the individual fishing effort, any level in the range $[0, E^{max}]$ can be an equilibrium, as the marginal resource rent from increasing fishing effort is nil.

Just why is this so? Go back to the original sole owner, who is joined by a second player, a non-cooperative player. The former sole owner now has no confidence in what it will now earn from the fishery in the future, and will, if rational, discount those future returns massively. The second player may well deplete the resource, reasons the former sole owner. If so, it is better to deplete the resource and enjoy the temporary economic benefits from so doing, before all is taken by the second player. Remember that, if the player's discount rate equals infinity, Bionomic Equilibrium appears to be optimal. The second player will come to exactly the same conclusions.[19]

This is, to say the least, a striking outcome. In the case of the static symmetric non-cooperative fishery game, Bionomic Equilibrium will be approached, only if $n = \infty$. In the dynamic symmetric game, the resource will be driven down to the Bionomic Equilibrium, if $n = 2$! Why the difference? In the static game, the net economic returns to a player are derived from sustainable harvests only. In the dynamic game, this constraint does not exist. The net economic returns from the fishery to a player are derived from any harvest that it may take, sustainable or unsustainable.

n-Player Game

Let us now extend the game to any number of players, $n \geq 2$. The dynamics of the stock is now represented by

$$\frac{dX}{dt} = F(X(t)) - \sum_{i=1}^{n} H_i(t). \tag{3.58}$$

Each player $i = 1, 2, \ldots, n$ solves the following problem:

$$\max_{E_i(t)} PV_i = \int_0^{+\infty} e^{-\delta t}(pqX(t) - c)E_i(t)dt$$

$$\text{s.t.} \quad \frac{dX}{dt} = F(X(t)) - qE_i(t)X(t) - \sum_{j \neq i} qE_j(t)X(t) \tag{3.59}$$

$$X(0) = X_0, X(t) \geq 0$$

$$0 \leq E_i(t) \leq E^{max}.$$

---

[19]What we have described is, of course, a manifestation of the Prisoner's Dilemma. Return to the discussion of non-cooperative games in Chap. 2.

Solving this problem for all players simultaneously yields the Feedback Nash equilibrium fishing effort.

$$E_i(t) = \begin{cases} 0 & \text{if } X(t) < X^{OA} \\ \alpha_i \frac{F(X^{OA})}{qX^{OA}} & \text{if } X(t) = X^{OA} \\ E^{\max} & \text{if } X(t) > X^{OA} \end{cases} \tag{3.60}$$

where $0 \leq \alpha_i \leq 1$, and $\sum_{i=1}^{n} \alpha_i = 1$.

As, in the case of two players, the strategy of each player corresponds to a most rapid approach to the bionomic equilibrium stock level, X^{OA}, in which resource rent is nil. Moreover, the Feedback Nash equilibrium is not unique as, in the Bionomic Equilibrium, any partition of the aggregate effort, $\frac{F(X^{OA})}{qX^{OA}}$, among the players is an equilibrium. This holds as no player has incentive to change its fishing effort, for any partition of the aggregate effort.

Adding more symmetric players to the game does not affect the equilibrium stock level, X^{OA}. However, it makes the path towards that equilibrium faster whenever the stock surpasses that level, as the sum of the maximum fishing effort capacity over all players increases.

Example 3.3 Symmetric players. Consider a dynamic game between two-symmetric players. Let the dynamics of the fish stock be described by a logistic function. Take the fishery parameters used in Example 3.1: $r = 1$, $k = 100$, $q = 0.004$, $p = 3$ and $c = 0.3$. Furthermore, assume that the maximum effort capacity is $E^{\max} = 200$, and the discount rate is $\delta = 0.05$.

With these parameters the Bionomic Equilibrium of both players is $X^{OA} = 25$. The Feedback Nash equilibrium fishing effort strategies is given by

$$E_i(t) = \begin{cases} 0 & \text{if } X(t) < 25 \\ 187.5\alpha_i & \text{if } X(t) = 25 \\ 200 & \text{if } X(t) > 25 \end{cases} \tag{3.61}$$

where $0 \leq \alpha_i \leq 1$, and $\sum_{i=1}^{2} \alpha_i = 1$.

If the stock is below the Bionomic Equilibrium, both players cease fishing as they incur in losses. If the stock is above that level, both experience positive profits and hence use their maximum fishing effort, $E^{\max} = 200$. When the stock is at the Bionomic Equilibrium level, the aggregate fishing effort is 187.5. Any division of this effort level between the players is a Nash equilibrium, as in the Bionomic Equilibrium players have zero resource rent irrespective of their fishing effort.

3.2.3 Asymmetric Game

We will now explore the case in which the players are asymmetric. We introduce asymmetry by assuming that the players are identical in all respects, except with regards to fishing effort costs.

Two-Player Game

Without loss of generality, it will be assumed that player 1 has lower fishing effort costs than does player 2. We thus have $c_1 < c_2$. The model that we have been presenting is, as we have stated, a dynamic version of the Gordon–Schaefer model. In that model, harvesting costs are sensitive to the size of the stock, $X(t)$. As will be recalled from our discussion of Eq. (3.51), we can express unit harvesting costs as $c(X) = \frac{c}{qX}$, where the sensitivity of harvesting costs to the size of X is obvious.

The implication is that, for any given level of $X(t)$, player 1's unit harvesting costs will be lower than those of player 2. The consequence of this, in turn, is that $X_1^* < X_2^{*}$[20] and $X_1^{OA} < X_2^{OA}$.

The problem of player i can be expressed as follows:

$$\max_{E_i(t)} PV_i\big(E_i(t), E_j(t)\big) = \int_0^{+\infty} e^{-\delta t}(pqX(t) - c_i)E_i(t)\mathrm{d}t$$

$$\text{s.t.} \quad \frac{\mathrm{d}X}{\mathrm{d}t} = F(X(t)) - qE_i(t)X(t) - qE_j(t)X(t) \qquad (3.62)$$

$$X(0) = X_0, \ X(t) \geq 0$$

$$0 \leq E_i(t) \leq E^{\max}.$$

Clark (1980) demonstrates that the Nash equilibrium fishing strategies of this problem, provided $E^{\max}$ is sufficiently large, are as follows:

The Feedback Nash equilibrium fishing strategies are given by

$$E_1(t) = \begin{cases} 0 & \text{if } X(t) < \min\big(X_1^*, X_2^{OA}\big) \\ \frac{F(X)}{qX} & \text{if } X(t) = \min\big(X_1^*, X_2^{OA}\big) \,, \\ E^{\max} & \text{if } X(t) > \min\big(X_1^*, X_2^{OA}\big) \end{cases} \qquad (3.63)$$

$$E_2(t) = \begin{cases} 0 & \text{if } X(t) \leq X_2^{OA} \\ E^{\max} & \text{if } X(t) > X_2^{OA} \,. \end{cases}$$

---

[20]The high-cost player, if it were sole owner, would have a greater incentive to invest in X than would the low-cost player, if it were sole owner.

The Feedback Nash equilibrium corresponds to a most rapid approach to $\min\left(X_1^*, X_2^{OA}\right)$ by player 1. Let us consider two possible cases:

Case 1: $X_1^* < X_2^{OA}$

This case is very straightforward. Player 1 will simply drive player 2 out of the fishery, with player 1 becoming in effect the sole owner of the resource. While not a particularly good outcome for player 2, the outcome is optimal in the sense that the resource will be exploited at the minimum cost and will be stabilized at the low-cost player's optimal biomass level.

Case 2: $X_1^* > X_2^{OA}$

In this case, the fishery resource will be driven down to player 2's Bionomic Equilibrium level. This is clearly sub-optimal, but player 2 will find that it is no longer in its best interests to remain in the fishery.[21,22]

n-Player Game

Consider now that the number of players is $n \geq 2$. Without loss of generality, let us order the players according to unit cost of effort: $c_1 < c_2 \leq c_3 \leq \cdots \leq c_n$. The problem of each player $i = 1, 2, \ldots, n$ is the following:

$$\max_{E_i(t)} PV_i = \int_0^{+\infty} e^{-\delta t}(pqX(t) - c_i)E_i(t)\mathrm{d}t$$

$$\text{s.t.} \quad \frac{\mathrm{d}X}{\mathrm{d}t} = F(X(t)) - qE_i(t)X(t) - \sum_{j \neq i} qE_j(t)X(t) \tag{3.64}$$

$$X(0) = X_0, X(t) \geq 0$$

$$0 \leq E_i(t) \leq E^{\max}.$$

The Feedback Nash equilibrium fishing strategies are now given by

$$E_1(t) = \begin{cases} 0 & \text{if } X(t) < \min\left(X_1^*, X_2^{OA}\right) \\ \frac{F(X)}{qX} & \text{if } X(t) = \min\left(X_1^*, X_2^{OA}\right) \\ E^{\max} & \text{if } X(t) > \min\left(X_1^*, X_2^{OA}\right) \end{cases}, \tag{3.65}$$

---

[21] Why is this so? Let it be assumed that both players exercise iron control over their respective fleets. At, X_2^{OA} the revenue to the player 2 fleet will just be sufficient to cover the fleet's opportunity costs. Thus, at X_2^{OA}, player 2 would seem, at first glance, to be on the margin of indifference between leaving the labour and produced capital in the fishery, or re-deploying the two to other parts of the economy. But, if player 2 remains in the fishery, the optimal strategy for player 1 would be to set $E_1 = E^{\max}$, thereby driving X below X_2^{OA}. Once X_2^{OA} is reached, player 2's economically rational policy will be to adopt the re-deploy option, and to do so with all possible speed, thus leaving the fishery to player 1.

[22] Additionally, a Case 3 could also be considered: $X_1^* = X_2^{OA}$. As in Case 2, the stock is driven to Bionomic Equilibrium of player 2, who does not remain in the fishery. The difference is that now that stock level is optimal as it coincides with the sole owner level of player 1.

$$E_j(t) = \begin{cases} 0 & \text{if } X(t) \leq X_j^{OA} \\ E^{\max} & \text{if } X(t) > X_j^{OA} \end{cases}, \quad j = 2, 3, \ldots, n.$$

As in the case of a two-player game, only the most efficient player remains active. All other players are driven out of the fishery.

Example 3.4 Asymmetric players. Consider a dynamic game between two asymmetric players, which differ in terms of their unit cost of effort: $c_1 = 0.3$ and $c_2 = 0.5$. As in Example 3.3, the dynamics is described by a logistic function and the model parameters are $r = 1$, $k = 100$, $q = 0.004$, $p = 3$, $E^{\max} = 200$ and $\delta = 0.05$.

With these parameters, the optimal stock level of player 1 as a sole owner is $X_1^* \approx 61.02$, whereas the Bionomic Equilibrium of player 2 is $X_2^{OA} \approx 41.67$. As $X_1^* \geq X_2^{OA}$, we are in Case 2 situation as presented above. Thus, the equilibrium fishing efforts are

$$E_1(t) = \begin{cases} 0 & \text{if } X(t) < 41.67 \\ 145.83 & \text{if } X(t) = 41.67 \\ 200 & \text{if } X(t) > 41.67 \end{cases}$$

$$E_2(t) = \begin{cases} 0 & \text{if } X(t) \leq 41.67 \\ 200 & \text{if } X(t) > 41.67 \end{cases}.$$

3.3 Comparison of Static and Dynamic Games

The two classical games presented in this chapter have been used to model fisheries in different contexts. The static game, by focusing on steady state, sustainable yield, payoffs is far, far simpler than the dynamic game. For that reason, the static game is widely used in theoretical studies, especially when combined with coalition formation, as will be shown in Chaps. 5 and 6. The dynamic game has the advantage of allowing the fishing effort to vary over time through feedback strategies, in which the fishing effort depends on the stock level. Thus, the dynamic game is usually preferred in empirical applications.

How do the outcomes of the two games compare? Both predict that non-cooperation will produce sub-optimal results, that cooperation does indeed matter. Both predict that asymmetry will mitigate the effects of non-cooperation and can in some cases lead to an optimum, in which the more efficient player drives the less efficient one(s) out of the fishery. The outcomes of the dynamic game are, however, much more dramatic than those of the static game. This is most apparent in the case of symmetric games.

The test of any theory, of any model, rests upon its predictive power. In those cases in which the fishery resource is a fast-growing one, the static model is likely to come into its own. In Chap. 2, case study on the anchovy resource shared by Peru and Chile, the two-player game is non-cooperative. The results are clearly sub-optimal, but not disastrously so—a static game type outcome. The resource is a very fast-growing one.[23] The case study to follow will be on a very slow-growing resource. In this case study, it will be seen that the dynamic symmetric non-cooperative game predicted brilliantly.

3.4 Case Study: The South Tasman Rise Trawl Fishery

This South Tasman Rise trawl fishery illustrates the possible consequences of a cooperative fishery game degenerating into a non-cooperative one. The South Tasman Rise is a continental shelf area straddling the Australian EEZ and the adjacent high seas. The key fishery resource in the area is an orange roughy one. The resource is technically a straddling stock, but there are only two key fishing states involved, Australia and New Zealand. The resource can thus be viewed as a de facto transboundary one.

Orange roughy is a deep sea demersal resource that is at once both high valued and vulnerable to over-exploitation. The vulnerability arises first from that the fact that the resource is normally harvested during the spawning phase, when an intense aggregation occurs,[24] and second from the fact that the resource is very long lived (up to 150 years) with a concomitant slow growth rate[25] (Munro et al. 2004, 51).

In 1997, the Australian trawl fishing industry found significant amounts of orange roughy in the adjacent high seas portion of the South Tasman Rise, which aroused the interest of the New Zealand fishing industry. The Australian resource managers acted sensibly. Seeing New Zealand as the only other state seriously interested in the resource, the Australians approached their New Zealand counterparts and proposed the establishment of a cooperative management regime for the vulnerable resource. This should have been a straightforward undertaking as the two coastal states were (and are) close neighbours, with similar cultural and historical backgrounds. Moreover, both had exemplary records of intra-EEZ fishery resource management (Munro et al. 2004).

The two coastal states did, in December 1997, enter into a cooperative management agreement for the 1998–1999 season. There was to be an overall quota of 2100 t, with 80% for the Australians and the remainder for the New Zealanders,

---

[23]Marine biologists commonly gauge whether a fishery resource is fast growing, slow growing or in between by referring to the resource's intrinsic growth rate—its maximum rate of growth, denoted by r. If $r = 0.50$, that, at a maximum, the annual growth rate of the resource biomass will be 50%. It is estimated that for the anchovy resource in question, we have $r = 1.8$ (William Cheung, Institute of Oceans and Fisheries, University of British Columbia, personal communication).

[24]This implies that harvesting costs are insensitive to the size of the biomass. In fisheries in which harvesting costs are sensitive to the size of the biomass, a declining biomass means rising harvest costs that act as a brake on further exploitation.

[25]The intrinsic growth rate of orange roughy is estimated to be $r = 0.025$ (Bjørndal and Munro 2012, 226).

with the start date to be 1 March 1998. The agreement was, however, flawed. There was a lack of clarity about fishing activities to be allowed, between December 1997 and the beginning of March 1998. Furthermore, there were no provisions for dealing with possible free riders.[26]

These flaws created a potential Prisoner's Dilemma type of situation, with respect to this high valued, vulnerable resource. If the Australians (New Zealanders) had absolute confidence in the probity of the New Zealanders (Australians) and had complete confidence that no free riders would appear on the scene, they would wait patiently until the start date. If they had no such confidences, the rational strategy would be to strike first. The New Zealanders had such confidences; the Australians did not. The Australians went fishing and it is alleged harvested 2000 t in January and February of 1998–95% of the quota. The New Zealanders were understandably outraged (Munro et al. ibid.).

There was, nonetheless, an agreement developed between the two for the 1999–2000 season. It was agreed to work once again towards a 2100 t quota, with a 80–20% split as before. The Australians promised to behave this time round, and did so. The New Zealanders did not. Having been the "fool", the "goat" in the 1998–1999 season, they were determined not to be so in the following season. The New Zealand share of the quota for 1999–2000 was 420 t. It is estimated that the New Zealanders harvested in excess of 1600 t during that season (Munro et al. ibid.).

In addition, free riders appeared in the form of four vessels, three flying the flag of South Africa and one the flag of Belize.[27] Diplomatic action was taken to remove the free riding four, but not until the four had taken up to 6200 t. It is estimated that the actual harvest in the 1999–2000 season may have been as high as 10,000 t—almost five times the agreed upon quota (Munro et al. ibid.).

The Australians and New Zealanders, having learned their lesson, did in 2000 develop a well thought out, foolproof cooperative resource management agreement, which did, among other things, have provisions for dealing effectively with free riders. One could think of there having been a set of repeated non-cooperative games leading ultimately back to a cooperative one.

There was but one problem with the new cooperative resource management agreement. It came too late. In the 2001–2002 season, the actual total harvest was below 190 t (Munro et al. ibid.). Australian government authorities, at that time, took the view that "current indicators suggest a low remaining biomass and low future yields" (cited in Munro et al. 2004, 54).

There were further developments. Under the cooperative resource management agreement, Australia and New Zealand agreed after the mid-2000s that there should be an outright harvest moratorium on South Tasman Rise orange roughy. The moratorium came into effect for the 2007–2008 season. It was agreed further that the moratorium should continue beyond the 2007–2008 season indefinitely, until such

---

[26]Recall that the resource was and is in the adjacent high seas portion of the South Tasman Rise.

[27]There is anecdotal evidence suggesting that the vessels flying the South African flag, at least, were in fact reflagged New Zealand vessels. The evidence, let it be stated, is strictly anecdotal. Hard evidence does not exist.

time that conditions warranted resumption of harvesting. At the time of writing, more than a decade later, the harvest moratorium remains in place (Australian Fishing Management Authority 2018).[28]

There are two conclusions that can be drawn from this case. The first arises from the fact that the South Tasman Rise orange roughy game was and up to this point is basically a two-player game (with some help from a few free riders), and from the fact that it is reasonable to conjecture that the players were and are close to being symmetric. The outcome is that predicted by the two-player symmetric dynamic non-cooperative game. The resource was driven down to a Bionomic Equilibrium type of outcome, with Bionomic Equilibrium in this case looking uncomfortably like commercial extinction.

The second conclusion is not that Australia and New Zealand should be held up for censure. It is rather that, if these two developed fishing states with excellent domestic fisheries management records could fall into the Prisoner's Dilemma hole, then no set of fishing states sharing a fishery resource is safe.

3.5 Policy Implications

The outcomes of the non-cooperative games presented in this chapter have important policy implications. The basic implication is that cooperation does matter, and matters particularly both when the number of players is large, and when the relevant fishery resource is a slow-growing one. If the fishery resource is a slow-growing one, the number of players does not need to be large for non-cooperation to lead to disastrous results.

In addition, the dynamic game, in particular, illustrates how the most efficient player finds it optimal to drive out the less efficient ones by harvesting the stock down to levels at which the less efficient players find it in their best interests to leave the fishery. This highlights how strategic interaction in a dynamic context can lead to entry deterrence, in which a dominant player keeps the stock at a sufficiently low level to avoid the entry of competitors.

---

[28]Further, Australia and New Zealand have, at the time of writing, plans to transfer the management of the fishery in the adjacent high seas to the South Pacific Regional Fisheries Management Organization. Recall that technically speaking the resource is a straddling one. Within the Australian Fisheries Zone component of the South Tasman Rise, no fishing of orange roughy is allowed (Professor Stephanie McWhinnie, University of Adelaide, personal communication).

References

Bjørndal, T., & Munro, G. (2012). *The Economics and management of world fisheries*. Oxford University Press.

Clark, C. W. (1976, 1990). *Mathematical bioeonomics: The optimal management of renewable resources* (1st, 2nd ed.). New York: Wiley.

Clark, C. W. (1980). Restricted access to common-property fishery resources: A game theoretic analysis. In P. Liu (Ed.), *Dynamic optimization and mathematical economics* (pp. 117–132). New York: Plenum Press.

Clark, C. W. (2010). *Mathematical bioeonomics: The mathematics of conservation* (3rd ed.). New York: Wiley.

Clark, C. W., & Munro, G. (1975). The economics of fishing and modern capital theory: A simplified approach. *Journal of Environmental Economics and Management, 2*, 92–106.

Clark, C. W., & Munro, G. (2017). Capital theory and the economics of fisheries: Implications for policy. *Marine Resource Economics, 32*, 123–142.

Clark, C. W., Clarke, F., & Munro, G. (1979). The optimal management renewable resource stocks: Problems of irreversible investment. *Econometrica, 47*, 25–47.

Gordon, H. S. (1954). The economic theory of a common property resource: The fishery. *Journal of Political Economy, 62*(2), 124–142.

Mesterton-Gibbons, M. (1993). Game-theoretic resource modeling. *Natural Resource Modeling, 7*, 93–147.

Munro, G. R., Van Houtte, A., & Willmann, R. (2004). *The conservation and management of shared fish stocks: Legal and economic aspects*. FAO fisheries technical report 465.

Chapter 4
Two-Player Cooperative Games

Abstract Chapter 3 discusses at length the theory of non-cooperative games, applied to fisheries management, and has as its fundamental conclusion that non-cooperation in the management of shared fishery resources carries with it the risk of serious resource mismanagement. Thus, cooperation in such management does indeed matter. Chapter 2 introduced the theory of cooperative games, applied to fisheries management, but did so only briefly. This chapter continues with the introduction of cooperative fishery games, doing so in greater detail. Having said this, the chapter restricts itself to two-player cooperative games, due to the sharp differences between two, and three or more, player cooperative games. While restrictive, it will be seen that two-player cooperative games have decided policy relevance to international transboundary fishery resources, as will be illustrated by several case studies. Cooperative fishery resource management will be a theme that will carry through all of the chapters to follow.

4.1 On Understanding Cooperative Games in Fisheries

Chapter 2 defines a cooperative game as one in which commitments (agreements, promises, threats) are binding and enforceable. The requirements for a stable cooperative game are thus demanding, to say the very least. Hence, it is necessary first to examine the conditions, which could lead to a cooperative agreement that appears, at least from the outset, to be binding. Then, it becomes necessary to find means of ensuring that the agreement continues to be binding through time. If success is achieved, it could lead to a first best outcome. In the case studies to follow, a cooperative fishery game will be discussed in which the first best outcome, or something very close to it, was achieved. If even only a second best, or third best, is achieved, it is, with very few exceptions, certain to be far better than the non-cooperative alternative.

In order for a cooperative game to be within the realm of possibility, two (obvious) prior conditions must be met. First, the players must be able to communicate effectively. In the Prisoner's Dilemma discussed in previous chapters, communication between the players is impossible, which means, in turn, that cooperation between the two will not hold. Second, the players must have the motivation to engage in what might be a difficult and costly process of establishing a cooperative resource management agreement. In international fisheries, the motivation for cooperation has, more often than not, come from the players witnessing the dire consequences of non-cooperation.

In this chapter, the discussion will be focused on two-player games exclusively. Denoting, as usual, the number of players as n, it can be stated that in non-cooperative games there is not much difference between $n = 2$ and $n > 2$ games. The $n > 2$ case can be seen as a generalization of the $n = 2$ one (Owen 2013, 207). We did, in fact, see this very clearly in Chap. 3.

In cooperative games, by way of contrast, there is a very significant difference between the two. In $n > 2$ cooperative games, coalitions become of paramount importance, compelling us to distinguish between sub-coalitions and the Grand Coalition. Many complications arise. In the $n = 2$ case, there is but one coalition that really counts, and life is much, much simpler (Owen 2013, ibid.). Examination of the complexities of $n > 2$ cooperative games will be postponed until Chap. 5. While confining the discussion in this chapter to $n = 2$ cooperative fishery games may seem to be very restrictive, Chap. 1 has revealed that two-player cooperative games can, in fact, carry us some distance in analysing the economic management of transboundary fishery resources.

Next, it is necessary to comment on so-called side payments, or transferable utility, as it will be referred to in later chapters. These will be seen to play a crucial role, particularly in $n > 2$ cooperative games. Side payments, or side payment like arrangements, can be thought of as transfers in monetary or non-monetary form, or side agreements, between or among the players. As it will be seen, side payments can serve to stabilize cooperative agreements and to make them more effective.[1]

For our purposes, we shall deem a two-player cooperative fishery game *without* side payments to be one in which the payoff to each player is determined solely by the harvests of that player's fleet in that player's EEZ. For example, suppose that the cooperative game involves the management of a transboundary fishery resource shared by two coastal states, 1 and 2. In the cooperative fishery game without side payments, the payoff to coastal state 1 (2) will be determined solely by the harvests of the 1 (2) fleet in the 1 (2) EEZ.

In the two-player cooperative game, the three basic conditions necessary for a stable solution to the game were set forth by von Neumann and Morgenstern (1944). The conditions are presented in Definition 4.1.

---

[1]Let us take a simple non-fisheries example to illustrate. Three neighbours have in the past all driven to work in their own cars. They enter into a cooperative carpooling arrangement to reduce expenses. The arrangement could be such that each neighbour drives her other two neighbours every third working day. They realize, however, that the first neigbour's car is much more suitable for the carpooling arrangement than the other two cars. It is agreed that only the first neighbour's car is to be used. To make this agreement stable, the second and third neighbours make side payments to the first neighbour, to cover two-thirds of her driving costs.

Definition 4.1 Denote two players, 1 and 2, and their payoffs, μ_1, μ_2.

Feasibility holds if the payoffs, μ_1, μ_2, belong to the set of all feasible payoffs to the game. Denote S as the feasible set of payoffs, thus $(\mu_1, \mu_2) \in S$.

Collective Rationality implies that there cannot exist an alternative solution that would make both players better off, or make one player better off without harming the other. In other words, the solution must be Pareto optimal. Denote σ as the total payoff from the two-player game, thus $\mu_1 + \mu_2 = \sigma$.

Individual Rationality holds if the players by cooperating will be at least as well off as they would be by not cooperating. Denote the payoffs to a non-cooperative game between the two as $\overline{\mu_1}, \overline{\mu_2}$. Thus, $\mu_1 > \overline{\mu_1}$ and $\mu_2 > \overline{\mu_2}$, respectively.

Feasibility, collective and individual rationality are three basic conditions necessary for a **stable solution** to a two-player game.

In the two-player cooperative game, the basic conditions necessary for a stable solution to the game basically say that the solution must exist, nothing should be leftover and there will not be a stable solution to the cooperative game unless both 1 and 2 are convinced that by cooperating they will be at least as well off as they would be by not cooperating. Altruism plays no role. It is reasonable to think of the alternative for both players as the solution payoffs to a non-cooperative game between the two. Those solution payoffs satisfying the three conditions can be seen, in the context of the two-player game, as constituting the Core of the game (see Bacharach 1976; Owen 2013).

The problem is that there are likely to be many sets of solution payoffs that satisfy the three conditions, with the consequence that we are faced with an indeterminacy. The problem of indeterminacy was resolved by Nash (1953), who put forward his bargaining solution. It is to that solution we now turn.

4.1.1 The Nash Bargaining Solution

The Nash bargaining solution is a concept for cooperative games. It is therefore to be distinguished from the Nash equilibrium presented in Chap. 2, which is directed to non-cooperative games. Under the Nash bargaining solution, players are assumed to determine, through negotiation, the point in the feasible set of payoffs upon which they will agree. It is further assumed that the players only care about their outcome of the bargaining and not about the process leading towards the agreed solution.

Definition 4.2 The Nash bargaining solution is the vector, $\mu \in \mathbf{R}^2$, that maximizes the Nash product $\max_{\mu_i} N(\mu) = (\mu_1 - \overline{\mu_1})(\mu_2 - \overline{\mu_2})$, s.t. $\mu_i > \overline{\mu_i}, \forall i = 1, 2$, where μ_i is the payoff to player i from the sharing of benefits from the Grand Coalition and $\overline{\mu_i}$ is the threat point or disagreement point that will occur if no agreement is formed.

The Nash bargaining solution is based upon five axioms or requirements:

(1) *individual rationality,*
(2) *Pareto optimality,*
(3) *independence from irrelevant alternatives* (IIA),
(4) *linear invariance* (LI) and
(5) *symmetry.*

Nash proved that there is one unique rule satisfying these requirements (Gravelle and Rees 1992; Owen 2013). The first requirement, the individual rationality, is defined in Definition 4.1 and ensures that the individual player has an incentive to join the solution. Nash designates $\overline{\mu_1}, \overline{\mu_2}$ as the Threat Point payoffs. These are to be seen as a measure of relative bargaining strength.[2]

The second requirement, Pareto optimality, also referred to as collective rationality (see Definition 4.1), ensures that there exists no feasible alternative solution, which would make at least one player better off without making any other player worse off. The third requirement, IIA, compares two games with same threat point, or disagreement point. One of the games is contained in the other. The axiom states that, if the solution to the larger game is included in the feasible set of the smaller game, then the two games must have the same outcome. If the feasible set, S, is enlarged, the solution to the new problem will either be the original solution or one of the new points in the set, not another point in the original, smaller set (Owen 2013, 188).

The fourth axiom, LI, also considers two games, where one game is formed by a positive affine transformation of the other game. These two games have exactly the same set of potential bargains and solutions. The disagreement points are the same and as are the preferences. The only difference is the numerical representation of the preferences. Hence, relabeling the utility functions re-labels the solution outcome in the exact same way. Finally, the fifth axiom, symmetry, implies that if the bargaining game is symmetric it implies identical preferences and identical circumstances, then the solution to the game must be symmetric too. If it is found that $\overline{\mu_1} = \overline{\mu_2}$, then it must be the case that $\mu_1 = \mu_2$ (Owen 2013).

One additional axiom that is sometimes added is the (6) *Independence with respect to linear transformations of the set of payoffs* can best be explained with an example. The payoffs in the fisheries games to be discussed will be expressed in monetary terms—net economic returns. In a particular case, we might want to express the payoffs in terms of US dollars. If we were to decide to transform these payoffs into Euros—linear transformation through the US dollar–Euro exchange rate—this would have no impact upon the set of solution payoffs.

---

[2]The basic idea is that the less a player has to lose from the attempt to cooperate not succeeding, the greater its bargaining power. Thus, the larger (smaller) $\overline{\mu_1}$ is in relation to $\overline{\mu_2}$ the greater (less) is 1's bargaining power with respect to 2.

4.2 Two-Player Cooperative Fishery Games Without Side Payments

Game theorist, Guillermo Owen, argues that the introduction of side payments (trans-ferable utility) to the theory of cooperative games simplifies the analysis enormously (Owen 2013, 355). These authors can only agree and will demonstrate this fact in the context of cooperative fishery games. Owen continues that in the real world, how-ever, side payments are, for reasons right or wrong, not always feasible. He does, as a consequence, find it necessary to devote a full chapter of his book on game theory to cooperative games without side payments (Owen 2013, Chap. 15).

In the real world of fisheries management, side payments and side payments like arrangements are beginning to appear more frequently. The acceptance of side pay-ments has, however, been a slow process. The very name side payments has in the past had unfortunate connotations to policymakers, suggesting bribes, "kickbacks", all in all quite unsavoury (Munro 2013). Economists have been undertaking a campaign among policymakers to convince them of the benefits of side payments.

Given this state of affairs, it is essential to be able to demonstrate that the theory of cooperative games has relevance and applicability in situations in which, for whatever reason, side payments are infeasible, or are willfully ignored. This then is the task before us.

The task shall be undertaken, not by talking in generalities, but rather by develop-ing a specific fisheries model, in which side payments are explicitly not employed. The model that we shall bring to bear is a dynamic model, in fact exactly the same model that is employed in Chap. 3, which we described as a dynamic version of the Gordon–Schaefer model (Gordon 1954).[3]

As we did on Chap. 3, let us start out with the sole owner perspective, to give us a benchmark. In Chap. 3, we saw the objective of the sole owner as being that of maximizing the present value of the net economic returns, resource rent, through time. Technically, given our assumptions, we were presented with a linear autonomous optimal control problem. We then talked about a "state" variable, $X(t)$, and a "control" variable. With regards to a "control" variable, it was maintained that we had a choice between $H(t)$ and $E(t)$. In Chap. 3, it was most convenient to let $E(t)$ be the "control" variable. In this chapter, convenience is best served by letting $H(t)$ play that role.

In Chap. 3, we expressed the sole owner's objective functional as

$$PV(E(t)) = \int_0^{+\infty} e^{-\delta t}(pqX(t) - c)E(t)dt. \tag{4.1}$$

We will now express the sole owner's objective functional as[4]:

---

[3] We find that, in our discussion of two-player cooperative games, there are no benefits to be gained from a static model version.

[4] Recall from Chap. 3 that we have in this model: $H(t) = qX(t)E(t)$.

$$PV = \int\limits_{0}^{+\infty} e^{-\delta t}(p - c(X))H(t)dt, \tag{4.2}$$

where $c(X) = \frac{c}{qX}$.

The sole owner's objective is to maximize the PV of resource rent from the fishery over time (Eq. 4.2), subject to exactly the same constraints set out in Chap. 3.[5] The optimal biomass level, X^*, will be given to us by exactly the same decision rule as we had in Chap. 3, namely[6]:

$$F'(X^*) - \frac{c'(X^*)F(X^*)}{p - c(X^*)} = \delta. \tag{4.3}$$

We are now prepared for the game. Let it be supposed that the sole owner is a coastal state. The sole owner now discovers that the resource is in fact shared with a neighbouring coastal state. The two coastal states we shall refer to as players 1 and 2. We shall assume further that the resource is wholly confined to the EEZs of 1 and 2, so that the shared fishery resource is a strictly transboundary one. It will be assumed further that the two coastal states are selling their harvests into a world market at a common price p, which is constant over time.

We shall now consider two sub-cases, one in which the two are playing a symmetric game, and one on which the two are playing an asymmetric game, by virtue of differing fishing effort costs just as in Chap. 3.

4.2.1 A Symmetric Cooperative Game

The two players are assumed to be identical in all respects. Let it be supposed that the players were initially playing competitively. After observing the costs of non-cooperation, and being able to communicate with one another, the two finally respond to the admonition of Article 63(1) of the 1982 UN Convention[7] to explore the possibility of managing the fishery resource cooperatively.

In so doing, we assume that they are not prepared to contemplate side payments. What will be found is that the refusal so to contemplate will have no negative consequences, simply because, in this case, side payments will have no useful role to play.

As in Chap. 3, let us denote the optimal biomass level, as perceived by players 1 and 2, as X_1^*; X_2^* respectively. With symmetry prevailing, we have $X_1^* = X_2^*$. The perceived optimal resource management programme of player 1 will be identical to

---

[5]See Eq. (3.50).

[6]The optimal "approach path" from $X(0)$ to X^*, by the way, is exactly the same as in Chap. 3, i.e. the most rapid (see Eq. 3.53).

[7]UN (1982), Article 63(1).

Fig. 4.1 A symmetric cooperative fishing game

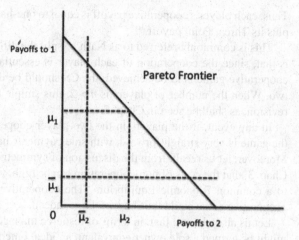

that of player 2. If the two players decide that cooperation is worthwhile,[8] they will, if rational, attempt to maximize the present value of the global resource rent through time, and then bargain over the division of the global resource rent. The game then is over that division.

Consider now Fig. 4.1. The 45° line is the Pareto Frontier. The sum of solution payoffs, μ_1, μ_2, at any one point on the Frontier is identical to the sum at any other point on the Frontier. A solution to the game is the Nash bargaining solution (see Definition 4.2).[9]

We need to say much more on the actual division of the net economic returns from the fishery, between players 1 and 2, arising from this Nash game. In order to do this, we must first present the concept of the cooperative surplus.

Definition 4.3 The cooperative surplus (CS) is defined as $CS = [(\mu_1 + \mu_2) - (\overline{\mu_1} + \overline{\mu_2})]$

The Nash bargaining solution payoffs, μ_1, μ_2, are then given by

$$\mu_1 = \left[\frac{CS}{2} + \overline{\mu_1} \right] \tag{4.4}$$

$$\mu_2 = \left[\frac{CS}{2} + \overline{\mu_2} \right] \tag{4.5}$$

---

[8] i.e. the Core of the game is non-empty.

[9] We have, in fact, seen this figure before. Return to Chap. 2 and consider Fig. 2.5.

Thus, each player's cooperative payoff is equal to one-half of the cooperative surplus plus its Threat Point payoff.[10]

This is commonly referred to as Nash's "split the difference" rule. The argument is that, since the cooperation of each player is essential if a stable solution to the cooperative game is to be achieved, the CS should be divided equally between the two. When the number of players is $n > 2$, this simple rule will require substantial revision, as shall be seen in Chap. 5.

In any event, if the players in the two-player cooperative game are symmetric, the game is very straightforward, with side payments having no useful role to play. Moreover, let us recall, from the discussion of symmetric non-cooperative games in Chap. 3, that the costs of non-cooperation can be high, with the players being driven to a common Bionomic Equilibrium. The cooperative surplus from a symmetric cooperative game could thus be very large indeed.

Let us also observe that, in terms of resource management, we end up with what might be termed a sole owner equivalent, an ideal outcome. See Eq. (4.3).

Symmetry, in real-world fisheries, is the exception, not the rule. Moreover, a cooperative fisheries game that is symmetric today, may not be so tomorrow, given the possibility of changing conditions. We must, therefore, enquire into asymmetric cooperative games. In so doing, it will be found that side payments definitely have a potential role. The question then, is it possible to have an asymmetric cooperative fisheries game with a stable solution in the absence of side payments? This question we must now explore.

4.2.2 An Asymmetric Cooperative Game

In this asymmetric game, it will be assumed that the two coastal states, players 1 and 2, are identical in all respects, *except* with regards to fishing effort costs, which we will denote by c_1 and c_2, respectively. It shall be assumed that $c_1 < c_2$. If we return to Chap. 3, we will be reminded, given our model, that for any level of X we shall have $c_1(X) < c_2(X)$, which in turn implies that, in contrast to the symmetric case, the perception of the optimal level of X will differ between 1 and 2. We shall have $X_1^* < X_2^*$.[11]

It is not immediately clear that there would be scope for cooperation that the Core of a potential cooperative game would in fact be other than empty. We know

---

[10]We can go a bit further. Each player's payoff is, in fact, equal to one-half of the total net economic return arising from the solution to the game. Since the game is perfectly symmetrical, the Threat Point payoffs, $\overline{\mu_1} = \overline{\mu_2}$, are equal to one another.

[11]Let us be a bit more specific on why this is so. Return to Eq. (4.3) and look at the second term on the LHS, which takes into account the impact of a marginal investment in X upon harvesting costs, referred to in the literature as the Marginal Stock Effect. For any given level of X, we shall have the following: $-\frac{c_2'(X)F(X)}{p-c_2(X)} > -\frac{c_1'(X)F(X)}{p-c_1(X)}$. The consequence of this is that 2 has a greater incentive to invest in X than does 1. Let us also add in passing that, of course, we have $X_1^{OA} < X_2^{OA}$.

from Chap. 3 that, if players 1 and 2 refused to cooperate, there would be three possibilities, namely, $X_1^* < X_2^{OA}$, $X_1^* = X_2^{OA}$ and $X_1^* > X_2^{OA}$. If either of the first two possibilities were to occur, there would be no basis for cooperation.[12]

Let it be supposed that the third possibility, $X_1^* > X_2^{OA}$, occurs. It is by far the most likely of the three. Now, there is clearly scope for cooperation.

One can see at once that side payments could play a role. Indeed, if side payments were feasible, global harvesting cost minimization, and thus global resource rent maximization would demand that all of the harvesting of the resource should be done by player 1. In such circumstances, one could think of player 2 importing the harvesting services of player 1. By assumption, however, players 1 and 2 are not prepared to contemplate side payments. Let us see what can be done.

In this asymmetric game without side payments, the payoff to player 1 will depend upon both player 1's share of the harvest through time and upon the resource management policy through time that is adopted. There is no assurance whatsoever that the resource management policy adopted will be the one that player 1 deems to be the optimum. What applies to player 1, applies with equal force to player 2.

In following the lead of Hnyilicza and Pindyck (1976), we first look at harvest shares.[13] Let us denote 1's harvest share as α, and 2's share simply as $(1 - \alpha)$. There is no necessary reason why these shares should be constant through time. Both the cases of α being constant over time and that of $\alpha = \alpha(t)$ will be considered.

With respect to the first step, let us start with the case of α being constant over time. We then, in effect, have a two stage game. Stage one, determine α; stage two determine the resource management policy through time.

As for stage one, what we can say right off is that with α constant through time, it is obvious that $0 < \alpha < 1$, if the individual rationality constraints are to be satisfied. Beyond this, we will simplify further by looking at the real world. In the real world, the determination of α is typically done without excessive negotiation, being usually done on the basis of some formula such as harvesting histories or zonal attachment—the amount of the resource to be found in EEZ_1 and EEZ_2, respectively.[14]

---

[12]No cooperative arrangement could satisfy player 1's individual rationality constraint.

[13]Hnyilicza and Pindyck were in fact looking at a decidedly non-fisheries problem—petroleum and OPEC, but their analysis remains entirely relevant to fisheries.

[14]An example is provided by the North Sea herring resource case study, discussed in Chap. 2. It will be recalled that the resource has been managed cooperatively by the EU and Norway. The two have harvest shares that are fixed over time, determined by zonal attachment. Scientists from ICES have determined that 71% of the resource is to be found in EU, and 29% in Norwegian waters. If we let the EU be 1, then $\alpha = 0.71$ (Bjørndal and Lindroos 2004), where zonal attachment or harvesting histories are not clear, a simple 50:50 rule is commonly adopted.

In turning now to stage two, let us be reminded that we have this stage, because $X_1^* \neq X_2^*$.[15] Hence, 1 and 2 must bargain over resource management and try to reach a compromise.[16] Denote the compromise optimal biomass level as X_{comp}^*. Once X_{comp}^* is determined, the optimal approach path from $X(0)$ to X_{comp}^* is the most rapid one.

To deal with this stage two of bargaining and the determination of X_{comp}^*, we proceed by introducing a bargaining parameter, $0 \leq \beta \leq 1$. If $\beta = 1$, the management preferences of 1 are wholly dominant. If $\beta = 0$, the management preferences of 2 are wholly dominant. If β lies between these extremes, we end up with a true compromise management programme, i.e. $X_1^* < X_{comp}^* < X_2^*$. The cooperative game is now about the determination of β.

The next step is to determine the Pareto Frontier. This is done by maximizing the weighted sum of PV_1 and PV_2, with the weights being given by the bargaining parameter.[17] We have:

$$\max PV = \beta PV_1 + (1.-\beta)PV_2. \tag{4.6}$$

Solve Eq. (4.6) for each and every β between 0 and 1. In so doing, we will have the Pareto Frontier in the space of realized payoffs. Given that the Core of the game is not empty,[18] the Nash bargaining solutions come into play (Definition 4.2), and thus leading to the determination of β.

If, for example, the resultant solution payoffs, μ_1, μ_2, were to be associated with $\beta = 3/4$, then the management preferences of both 1 and 2 would count, but the management preferences of 1 would be given three times the weight of those of 2.

Consider now Fig. 4.2, which shows the Pareto Frontier for a *given* α. We must first observe that for every other α, $0 < \alpha < 1$, there would be a separate Pareto Frontier. Note that the Frontier looks rather different from the one associated with a symmetric game.[19]

Once again, the Core of the game is shown by that part of the Pareto Frontier segmented by the dashed lines emanating from the Threat Point payoffs. This leads us to note that, while player 1 would clearly be best off if $\beta = 1$, such an outcome is not feasible. The point on the Frontier at which $\beta = 1$ lies outside the Core. Similarly, player 2 would be best off if $\beta = 0$, but such an outcome is also infeasible.

With β determined, the resultant solution to Eq. (4.6) leads to X_{comp}^*, which is given by a resource investment decision rule, comparable to Eq. (4.3). That equation,

---

[15]But what in fact does player 1's resource investment decision rule look like once α is determined? It can be shown that player 1's resource investment decision rule takes the following form: $F'(X_1^*) - \frac{\alpha c_1'(X_1^*)F(X_1^*)}{\alpha(p-c_1(X^*))} = \delta$, which, of course, reduces to $F'(X_1^*) - \frac{c_1'(X_1^*)F(X_1^*)}{(p-c_1(X^*))} = \delta$. What holds true for player 1 holds true for player 2 (Munro 1979).

[16]Let it be recalled from the model that we are using that there is a key assumption, namely, that the produced and human capital employed in both the player 1 and player 2 shares of the fishery are perfectly malleable.

[17]See Hnyilicza and Pindyck (1976), Munro (1979).

[18]If the Core is empty, then the attempt to establish a cooperative resource management programme will be unsuccessful, and the players will revert to non-cooperation.

[19]See Fig. 4.1.

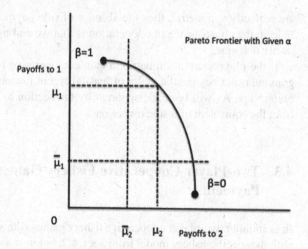

Fig. 4.2 An asymmetric cooperative fishing game with differences in fishing costs and no side payments

however, contains an exceedingly complex weighted Marginal Stock Effect, with the weights, being provided by α, $(1 - \alpha)$, β, $(1 - \beta)$.[20]

As a further example, let it be supposed that $X^*_{comp} = 1$ million tonnes (MT), and that the sustainable harvest associated with that biomass is 100,000 t per period. Suppose in addition that $\alpha = 0.70$; $(1 - \alpha) = 0.3$ Finally, suppose that $x(0) = 500,000$ t.

With $X(0) < X^*_{comp}$, the optimal policy is to maintain a harvest moratorium until $X(t) = X^*_{comp}$.

Suppose as well that scientists predict that the goal will be achieved in three years. 1's prescribed optimal harvest programme is, therefore, to do no harvesting for three years, and then harvest 70,000 t per period forever.[21] μ_1 will be equal to the PV of the net economic returns arising from this harvest programme, given p, $c_1(X)$ and δ. μ_2 is determined in a similar fashion.

Next, let us relax the assumption that α is constant over time. With α time-variant, the determination of α becomes part of a single bargaining package.[22] Munro (1979) demonstrates that, with side payments not being allowed, a solution to the cooperative game is achievable. The solution is, however, very cumbersome and very complex. So cumbersome and complex is it, and thus so lacking in policy relevance is it, that we should spend no further time on it.[23] We have to conclude that an asymmetric cooperative fishery game without side payments to be of policy relevance, must be one in which the harvest shares are determined on the basis of a straightforward formula.

The overall conclusion, in the context of fisheries, is that the two-player cooperative game analysis is of value, if side payments are disallowed. If the two players

---

[20]See Munro (1979, 365) for details.

[21]i.e. 70% of 100,000 t.

[22]With $\alpha = \alpha(t)$, α becomes a control variable in our optimal control problem.

[23]For verification, see Munro (1979, 364–367).

are perfectly symmetric, then the absence of side payments creates no hindrances. Indeed, the consequence of cooperation is that we end up with equivalent of a sole owner outcome.

If the players are asymmetric, a stable cooperative resource management programme is not beyond the realm of feasibility, but it cannot be seen as the *optimum optimorum*. As will be made apparent by the section to follow, the outcome is far from the equivalent of a sole owner one.

4.3 Two-Player Cooperative Fishery Games with Side Payments

In examining two-player cooperative fishery games with side payments, we continue with the specific fishery model from Sect. 4.2, but now assume that the players overcome their objections to side payments. With side payments allowed, the objective becomes that of maximizing the global net economic returns from the fishery through time, and then bargaining over the division of these net economic returns between 1 and 2.

With regards to the symmetric cooperative game, nothing changes. With regards to the asymmetric cooperative game (Sect. 4.2.2) a great deal changes. Return to Eq. (4.6). Maximizing the global returns from the fishery means giving equal weight to PV_1 and PV_2 (set $\beta = 1/2$), and then "letting the chips fall where they may".

In so doing, there are no difficulties in allowing α to be function of time—to the contrary. Part of "letting the chips fall where they may" involves optimizing with respect to $\alpha(t)$.[24] As Munro (1979) demonstrates, this drives us to the not unexpected conclusion that the optimal $\alpha(t)$ is $\alpha(t) = 1$, for all t, $0 < t \leq \infty$. In other words, the low-cost coastal state, player 1, is to do all of the harvesting, and is to do so until the end of time. It means as well, of course, that $X^*_{comp} \equiv X^*_1$. 2's harvesting costs are simply irrelevant. 1, and 1 alone, determines the optimal resource management policy.

Two comments are immediately required, both obvious. The first is that this is clearly optimal, if one's objective is to maximize the net economic returns from the fishery through time. If the 2 fleet does any of the harvesting, the net economic returns from the fishery will not be maximized. The second is that the policy is feasible, if and only if side payments can be employed.

In Chap. 1, reference was made to the *Compensation Principle*. The Principle states that where there are differences in management objectives between/among the players, it is all but inevitable that one player places a higher value on the fishery resource than does (do) the other(s). Optimal policy calls for allowing that player to dominate the resource management regime, and then compensate the other(s) through the use of side payments. Here one can see the Principle at work. The low-cost player 1 obviously places a higher value on the resource than does high

---

[24]See no. 20.

Fig. 4.3 An asymmetric cooperative fishing game with differences in fishing costs and no side payments. **A** Pareto Frontier (with given α) without side payments. **B** Pareto Frontier with side payments

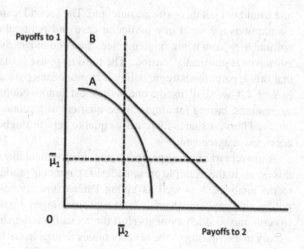

cost 2. The optimal policy described brings with it the complete dominance of 1's management preferences. Indeed, all of the harvesting is undertaken by 1. 1 then compensates 2 through side payments.

Consider Fig. 4.3, which shows the Pareto Frontier without side payments, and the Pareto Frontier with side payments. The Pareto Frontier with side payment lies everywhere above the Pareto Frontier without side payment. Why? If there are no side payments, 2 must do some of the harvesting for there to be a stable solution to the cooperative game, which in turn means that the global net economic returns from the fishery will not be maximized.

The Pareto Frontier with side payments looks exactly the same as we would have, if the players were symmetric.[25] No surprise here. Indeed, we have once again the equivalent of a sole owner outcome.

The Nash bargaining solution payoffs to 1 and 2 will be determined by the same "split the difference" rule that applied in the case of a symmetric cooperative game. See Eqs. (4.4) and (4.5).[26] The results with side payments allowed are neat and straightforward. There is, however, an apparently fundamental objection that can be raised against this policy. It is surely inconceivable, one can argue, that player 2 would find it politically tolerable to have its fleet's share of the harvest reduced to zero for ever and ever.

There are three responses to this objection. The first is that this seemingly extreme outcome is due to the linearity of our economic model of the fishery. If, for example, the supply of effort functions in 1 and 2 was found to exhibit finite price elasticity,

---

[25] See Fig. 4.1 again.

[26] There is an important qualification, however. In the case of the symmetric cooperative game, the players split the net economic return arising from the game equally. In the asymmetric game with side payments case, the two will divide the cooperative surplus equally, but not the net economic return arising from the game. Return to Chap. 3, which makes it clear that $\overline{\mu_1} > \overline{\mu_2}$. The low-cost player has more bargaining power than does its high-cost fellow player.

one would not get this extreme outcome. The second response is that, if the linearity assumptions are valid in a particular case and political considerations prevent the optimal $\alpha(t)$ from being implemented, then one moves as close to the *optimum optimorum* as is politically feasible. The third response is that history has demonstrated that this apparently extreme policy is by no means politically infeasible in all cases. In Sect. 4.4, we shall discuss one of the most famous cooperative fishery management agreements, lasting for almost three-quarters of a century, in which a set of players received harvest shares identically equal to zero throughout the entire history of the cooperative agreement.

A few remarks on side payments and side payment like arrangements are in order at this point. In the example presented, side payments produce a massive simplification to the analysis,[27] as well as giving Pareto superior results. In terms of practical policy, side payments have at least two advantages. First, which should be pleasing to economists, side payments shift the focus from the sharing of the harvests in the fishery to the sharing of the net economics returns from the fishery through time. The two are not always the same. Second, cooperative fishery games are about bargaining. In any bargaining situation, it is eminently desirable, for obvious reasons, to have the scope for bargaining be as broad as possible. Side payments serve this purpose. If players in a cooperative fishery game refuse to consider side payments, they are imposing upon themselves a constraint, which may be severe and which could doom the prospects for a stable solution to the game over time.

With the discussion of side payments complete, let us return to the discussion of asymmetric games in Sect. 4.2.2. Differences in Marginal Stock Effects led to differences in perceived optimal resource management programmes, between 1 and 2. These differences, in turn, arose because our model rests upon the Gordon–Schaefer model. It is well known, however, that there are fisheries for which this model, in particular the Schaefer component, is not strictly relevant.

As a consequence of the underlying Schaefer model, we have the following harvest production function, which was discussed in Chap. 3:

$$H(t) = qE(t)X(t) \qquad (4.7)$$

Properly, the equation should be expressed as

$$H(t) = qE^{\tau}(t)X^{\omega}(t) \qquad (4.7a)$$

where τ, ω are constants.

In the Gordon–Schaefer model, we have $\tau = \omega = 1$. The assumption that $\omega = 1$ arises from Schaefer's assumption that the fishery resource is always uniformly spread out through the relevant aquatic environment.[28] The assumption that $\omega = 1$

[27] All of which serves to support Owen's claim about the simplifying effects of introducing side payments (Owen 2013, 355).

[28] See Bjørndal and Munro (2012).

leads, in turn, to the economic result in the Gordon–Schaefer model that harvesting costs are sensitive to the size of $X(t)$.

There are several fisheries for which the aforementioned Schaefer assumption does not hold. This is particularly true of fishery resources characterized by intense schooling, examples of which we shall encounter in Sect. 4.4. Return to Eq. (4.7a). In such fisheries, it will be found that $\omega << 1$. In the extreme case, we have $\omega = 0$.

If $\omega = 0$, then, as the reader can verify, it will be found that $c_1'(X) = c_2'(X) = 0$. The consequence is that resource investment decision rules for players 1 and 2 become

$$F'(X_1^*) = \delta \qquad\qquad (4.8)$$

$$F'(X_2^*) = \delta. \qquad\qquad (4.9)$$

Thus, $X_1^* = X_2^*$.

In other words, the perceived optimal resource management programmes of players 1 and 2 will be identical.

What about side payments, will they be relevant? Side payments will indeed still be entirely relevant, for the simple reason that the net economic returns from the fishery will be maximized, if and only if all of the harvesting is done through time by the 1 fleet.

4.4 Case Studies

In this section, a few case studies shall be reviewed in brief with the purpose of illustrating a number of key points made in the previous sections. The case studies range from the Barents Sea, the Pacific Northwest of North America and the North Pacific.

4.4.1 Arcto-Norwegian Cod Fishery

The Arcto-Norwegian (or Northeast Atlantic) cod fishery is one of the largest and most valuable fisheries in the North Atlantic (Bjørndal and Lindroos 2012), being shared primarily by Norway and Russia in the Barents Sea. It is not a purely transboundary fishery, in that other states, e.g. EU and Iceland, are granted limited access to the fishery.[29] Let the resource be referred to as an almost transboundary resource.

At the end of the Second World War, the Arcto-Norwegian cod spawning biomass was estimated to be in excess of 4 MT. By the early 1970s, the biomass had been reduced by approximately one-half. Over that period annual harvests had in a few years exceeded 1 MT (Bjørndal and Lindroos 2012). This sounds similar to the

---

[29]There exists a high seas enclave in the Barents Sea, the Barents Sea Loophole.

experience of North Sea herring. During much of the Second World War, the Barents Sea was an exceedingly dangerous area in which to undertake fishing operations, so that it is no surprise that the resource was exceptionally large at the end of the war.

Cod is a much less vulnerable resource than herring, so that, with the decline of the resource, no one talked seriously about the resource being driven to extinction. Nonetheless, in 1975, Norway and the Soviet Union, anticipating the coming of the EEZ regime, and recognizing the benefits of cooperation, established a joint fishery commission. The Norwegians and Soviets agreed that, upon the implementation of their respective EEZs, they would through the commission manage cooperatively the Arcto-Norwegian cod stock, along with several other Barents Sea fishery resources. The EEZs were in fact implemented in 1977 (Armstrong and Flaaten 1991).[30]

Under the cooperative Norway–Soviet/Russian arrangement, annual TACs for the cod and related resources are established, based upon advice from ICES. With respect to the cod, there is a 15% set aside for third countries (Bjørndal and Lindroos 2012). The TACs for cod (outside of the set aside) and other fishery resources covered under the cooperative arrangement were and are to be divided on a 50:50 basis between the two.[31]

There are no explicit side payments between the Norwegians and the Soviets/Russians. Having said this, there have been and are side payment like arrangements. There are first allowances for quota swaps. Norway, for example, has typically taken in excess of 50% of the Norwegian–Soviet/Russian cod quota, in return for giving up part of its shares for other fishery resources (Armstrong 1994). As a further side payment like arrangement, a substantial part of the Soviet/Russian cod quota has been, and is, taken in the Norwegian EEZ (Bjørndal and Lindroos 2012). This makes eminently good economic (and biological) sense, as the mature cod is to be found primarily in the Norwegian EEZ (Armstrong 1994).

The side payments like arrangements are Pareto improving, thus serving to strengthen the cooperative arrangement. This is obvious with respect to the taking of the Soviet/Russian quota in the Norwegian zone. The quota swaps reflect different relative valuations, between the two coastal states of the fish resources in question. The swaps serve to make both Norwegians and Russians better off—almost a definition of a Pareto improvement.

The cooperative management regime has been far from perfect. The argument has been made, for example, that the agreed upon TAC has been exceeded several times. Whether this is due to Norwegian–Russian inadequate intra-EEZ management or outright cheating remains to be seen (Bjørndal and Lindroos 2012). Also the players are asymmetric. The management regime does not reflect that asymmetry.[32]

---

[30] One striking aspect of this agreement is that the Cold War was very much underway during the mid-1970s. Norway, at that time, was a member in good standing of NATO.

[31] Here is an example of α being determined on the basis of a simple formula.

[32] An early study of Arcto-Norwegian cod did attempt to examine the asymmetry between the two players. The study concludes that both the Soviet discount rate and harvesting costs were then lower than those of the Norwegians. The study also concludes that the two players' perception of the optimal biomass was not that different. It will come as no surprise to be informed that the lower discount rate, in of and by itself, would lead to the Soviets' perceived optimal biomass being higher

Be that as it may, while it is difficult to come up with exact measures of the cooperative surplus, the fact remains that the cooperative regime has been operating successfully for over 40 years, a fact made more remarkable by the political upheavals that have occurred over that period of time in one of the players (Soviet Union/Russia). One can safely surmise that both players see their cooperative payoffs to be well in excess of their Threat Point payoffs.[33]

4.4.2 Pacific Salmon Treaty of 1985

Throughout the latter part of the nineteenth century and most of the twentieth century, the dominant fishery resource for the commercial fisheries of the American states of Oregon, Washington and Alaska and the Canadian province of British Columbia was Pacific salmon, consisting of five sub-species. Pacific salmons are anadromous resources, being produced in freshwater habitats, migrating to the ocean and then returning to their freshwater habitats to spawn and die. In their migration, it is inevitable that some American produced salmon will be "intercepted", i.e. harvested, by Canadian fishers, and that some Canadian produced salmon will be "intercepted" by American fishers. The resource is thus very much a shared one, a transboundary resource.[34]

The Americans and Canadians began serious cooperative management of the Pacific salmon in the late 1930s, focusing on the Canadian Fraser River, south of Vancouver, the most important salmon river in the Western hemisphere.[35] This proved to be inadequate over time, leading the two coastal states to expand the cooperative management of Pacific salmon from northern California to the Gulf of Alaska, resulting in the Canada—United States Pacific Salmon Treaty of 1985 (Treaty 2009).

The post 1985 Canada–US Pacific salmon fishery game was and is not a strictly two-player game. While Canada can be seen as a single player, the United States is actually an informal coalition consisting of the states of Washington/Oregon, the state of Alaska, the Treaty Indian Tribes and the US federal government (Miller and Munro 2004).[36] Furthermore, the case involves an issue to be discussed in a later

---

than that of the Norwegians (see Munro 1979). Cod is resource in which one could expect harvesting costs to be sensitive to the size of the resource. The Soviets lower harvesting costs, in of and by themselves, would thus lead to the Soviets' perceived optimal biomass being *lower* than that of the Norwegians. Hence, the two sources of asymmetry worked in offsetting directions (Armstrong and Flaaten 1991).

[33] The aforementioned study attempts to measure the then cooperative surplus. It concludes that the cooperative surplus was then massive (Armstrong and Flaaten 1991).

[34] Some salmon migrate well beyond the American and Canadian EEZs. Under international treaty law, however, high seas harvesting of the Pacific salmon is forbidden (UN 1982, Article 66).

[35] Some of the returning Fraser River system salmon pass through American waters.

[36] In Canada, the management of fishery resources within the Canadian EEZs is the sole prerogative of the Canadian federal government, whereas in the United States the management of such resources

chapter, namely, that of uncertainty. Nonetheless, Pacific salmon serves as a useful case study at this point.

During its initial stages, the Treaty suffered from a lack of resiliency and the complete absence of side payments. The two are, of course, linked, because the absence of side payments exacerbated the resiliency problem.

As has been seen, central to any cooperative agreement is the "fair" division of the economic returns from that agreement. In the Treaty, this is expressed in what has come to be called the equity clause in which it is stated that each Party shall conduct its fisheries and its enhancement programmes so as to: "—provide for each Party to receive benefits equivalent to the production of salmon originating in its waters" (Treaty 2009, Article III). Measuring such benefits is, of course, a formidable task. In practice, in the early years of the Treaty the equity provision translated into a "fish for fish" rule. In the early 1990s, the then Canadian Minister of Fisheries and Oceans was quoted as saying the provision is designed to give each coastal state "the opportunity to harvest fish produced in its own rivers, or failing that, to harvest an equal amount of the other nation's fish" (cited in Miller and Munro 2004, 384). The possibility of side payments was deemed to be unworthy of consideration.

The Treaty calls for the establishment of a Pacific Salmon Commission to govern the overall harvest and allocation of the salmon stocks exploited jointly by the two coastal states, with frequent re-negotiations being required.[37] Initially, it was thought that the Fraser River of Canada and the Columbia River of the United States constituted the heart of the Treaty. The salmon fisheries of Alaska and northern British Columbia were seen as a "side show". It was believed by both coastal states that the American harvesting of Fraser River salmon was roughly offset by Canadian harvesting of Columbia River salmon (Miller and Munro 2004).

At the time that the Treaty came into force in 1985, a climatic regime shift was underway, a regime shift not properly recognized by the Treaty negotiators. The climatic regime shift had a most beneficial effect on Alaskan salmon stocks, leading to an almost tenfold increase in Alaskan salmon harvests between the mid-1970s and the early 1990s. The climate regime shift had the opposite effect on salmon stocks off southern British Columbia and off Washington, Oregon and northern California (Miller and Munro 2004).

The Alaskan harvesting of their Pacific salmon abundance had the consequence of increasing Alaskan interception of salmon produced in northern British Columbia. The increased Alaskan interception of British Columbia salmon could not be offset by increased Canadian harvesting of the beleaguered Columbia River stocks. Canada felt that the equity provision was being violated. The Alaskans were quite unwilling to reduce their harvesting, seeing this as an uncompensated cost to them. Indeed, to the Alaskans, the Treaty now seemed to be without benefit to them. In effect, the

---

within the American EEZs is shared between the US federal government and states bordering the oceans, e.g. Alaska.

[37]Each coastal state is to appoint three voting commissioners to the Commission. The three US commissioners represent Alaska, Washington/Oregon and the Treaty Indian Nations. There is no voting commissioner representing the US federal government (Miller and Munro 2004).

Alaskan individual rationality constraint was no longer being satisfied (Miller and Munro 2004).

By 1993, the situation had deteriorated to such an extent that the Pacific Salmon Commission was unable to agree on a new set of fishing regimes. While the Treaty was still in effect legally, it was in fact paralysed. The Treaty had been subject to an unpredictable environmental shock and was revealed to have lacked the resiliency to withstand the shock.

The cooperative fishery game degenerated into a competitive one. A "fish war" re-emerged, with Canada in particular adopting an "aggressive" fishing policy with the hope of forcing the Americans back to the bargaining table. This inflicted damage not just on American stocks, but on Canadian ones as well, providing a wonderful example of the Prisoner's Dilemma at work. The competitive game lasted for six dangerous years (Miller et al. 2001).

By 1999, the two sides, recognizing the increasingly damaging of the consequences of the competitive fisher game, came together to negotiate a "patch up" agreement. There was a shift from ensuring a balance of interceptions to ensuring the sustainability of the stocks—abundance-based management. This provided protection for the weakened stocks of southern British Columbia and the Columbia River system, while ensuring the Alaskans, with their abundant stocks, that they would not face stringent reductions in their harvests (Miller and Munro 2004).

In addition, the concept of side payments was introduced. Endowment funds for research and restoration in both the north and the south were made a part of the agreement. Both coastal states would benefit. While Canada was to make a contribution to the funds, its contribution was to be miniscule in comparison to the American contribution, and as such could be seen as an implicit side payment from the US to Canada (Miller et al. 2001). Needless to say, the term "side payment" appears nowhere in the agreement.

Over the following decade, the Treaty was re-negotiated, with the re-negotiated treaty coming into place in early 2009 (Treaty 2009). Interestingly, the revised treaty contains a more explicit side payment, although once again the term is avoided. One of the Pacific salmon sub-species of great concern at that time was the sub-species popularly known as Chinook. In order to provide support for the Columbia River Chinook, Canada was called upon to reduce its interception of the fish off the west coast of Vancouver Island, which would require the Canadian government to provide compensation to the affected fishers. Under the 2009 Treaty, the United States is to assist Canada in this purpose by granting to Canada US$30 million over a two-year period (Treaty 2009, Sect. 3.4). This is side payment plain and simple. While the side payment was not all that large, the precedent was clearly set.[38]

---

[38]What influence did the writings of academics have in the introduction of the precedent? This is very difficult to say. Interestingly, however, at the time that the 1999 "patch up" agreement was being negotiated, one of the authors (Munro) was a part of a small delegation that met with the Canadian negotiating team. A key point that the small delegation tried to make to the team was that a good part of the problem could be ascribed to the fact that the scope for bargaining ("fish for fish" rule) in the Treaty was far too narrow. Side payments, the small group stressed, serve to broaden the scope for bargaining.

The Treaty has remained in place and has worked reasonably well. It is being re-negotiated at the time of writing.

4.4.3 North Pacific Fur Seal Treaty

This is an historical case, but it is one that still has value, because it stands as one of the most successful undertakings in the cooperative management of fishery resources (Barrett 2003). It involved four players, but the four players formed two sub-coalitions, both so tight, that the case can be considered within a two-player game context.

During the second half of the nineteenth century, there developed a fur seal fishery in the North Pacific for the fashion trade. The fishery involved four participants (players), Canada, Japan, Russia and the United States. The fishery game at that time was non-cooperative, with a severe type of Prisoner's Dilemma type of outcome. The seal resource proved to be highly vulnerable to over-exploitation, to the extent that by the early years of the twentieth century there were fears that the resource could be driven to extinction (Bjørndal and Munro 2012, 195).

In response to the threat, the four came together and negotiated the North Pacific Fur Seal Treaty—the Convention for the Preservation of Fur Seals—in 1911 (Bjørndal and Munro ibid.). The players in the cooperative fishery game were asymmetric, with two distinct sub-coalitions emerging. The Americans and the Russians harvested the seals on land, on the Pribilof Islands, between Alaska and Russia. The Canadians and the Japanese harvested the seals at sea—hence the two distinct sub-coalitions. The harvesting costs of the Americans and the Russians were, not surprisingly, lower than those of the Canadians and the Japanese. Furthermore, due to better handling, the Americans and the Russians produced a better seal skin product than did the other two (FAO 1992, 45).

It is obvious that the Americans and the Russians placed a greater value on the seal resource than did the Canadians and the Japanese. From what has been said, ideally the US–Russia sub-coalition should have dominated the management of the resource. Indeed, that sub-coalition should have done all of the harvesting of the resource. That is exactly what happened. Under the terms of the Treaty, the Canadian and Japanese fleets were to be each given harvest allocations equal to zero, each and every year. In return, Canada and Japan were to receive certain percentages of the sealskins harvested by the Americans and Russians each and every year,[39] which were explicit side payments. The management of the resource, of the fishery, was of course entirely under the control of the Americans and Russians. This was the *Compensation Principle* at work.

---

[39]For example, in the post-World War II era of the Treaty, Canada and Japan each received 15% of the annual sealskin harvest from the Americans and Russians (FAO 1992, 45).

The Treaty remained in force until 1941. It was suspended during the remainder of the Second World War, was revived in the 1950s and remained in force until 1984 (Bjørndal and Munro ibid.).

The Treaty was unquestionably Pareto improving, in that it led to minimizing harvesting costs and to maximizing the prices received for the sealskins. The Treaty had resource conservation consequences as well. The FAO reports that the North Pacific seal herd is estimated to have numbered about 125,000 in 1911, at the time that the Treaty was signed. It is estimated that, at the time of the suspension of the Treaty in 1941, the seal herd numbered about 2,300,000—an 18-fold increase. It had paid the exploiters of the resource to invest heavily in the resource, this form of natural capital, over the 30-year period (FAO 1992, no. 23). If this was not a first best outcome, it was very close to it.

4.5 Policy Implications

In terms of policy issues, this chapter, focusing on two-player games, has been restricted to the management of the transboundary international fishery resources. This is a fishery management problem much simpler than those to be encountered in coming chapters. The policy implications are quite straightforward. First, the key conclusion of Chap. 3 is strengthened. In the management of transboundary fishery resource cooperation, with few exceptions, does matter. The Prisoner's Dilemma is very much a reality. This was shown to be strikingly the case with North Pacific seals. Non-cooperative management in these instances promised destruction of the resources. With regards to Pacific salmon, the players learned through experience that non-cooperation is costly. Even in the case of Arcto-Norwegian cod, where the resource was not threatened, the players perceived, correctly, that cooperation in resource management pays handsomely.

Second, experience has shown that long-term stable, binding, cooperative management regimes are, in fact, feasible in the case of transboundary fishery resources. Arcto-Norwegian cod illustrates this fact. Undoubtedly, the relatively small number of players helps.

What then leads to resource management cooperation, beyond communication between/among the players? The clear recognition by all players that a potential significant cooperative surplus exists is obviously essential. In the case studies explored, this recognition has been seen to arise in many instances, because of experience with the dire consequences of non-cooperative management.

Next, experience demonstrates that the satisfying of the individual rationality constraint does indeed matter. One of the factors leading to the seizing up of the Canada-US Pacific Salmon Treaty in the early 1990s was that this constraint was, for a time, allowed to remain unsatisfied for one of the players.

The experience reveals that side payments ("negotiation facilitators") are, one must concede, not always necessary. That being said, if players refuse to consider them, they do so at their cost. Side payments can result in larger economic payoffs for the players (larger cooperative surplus) and can, importantly, broaden the scope for bargaining. This can be seen in the case of Arcto-Norwegian cod, and could be seen dramatically in the case of North Pacific fur seals. With regards to Pacific salmon, another factor leading to the dangerous breakdown in the early 1990s was the narrowness of the scope for bargaining, exacerbated by the refusal even to consider side payments. The lesson was eventually learned.

There is one important policy implication brought out by the case studies that is to be explored in detail in later chapters. This relates to resiliency. All cooperative resource management arrangements are at risk of unpredictable shocks over time, which can be environmental, economic or political. If these arrangements lack resilience, the ability to withstand such shocks, the arrangements, while initially stable, can disintegrate turning a cooperative resource management arrangement into a competitive one, with all that implies. In two of the cases examined, such shocks arose: Arcto-Norwegian cod and Pacific salmon. The Arcto-Norwegian cod cooperative management regime displayed the requisite resilience; the Pacific salmon one did not.

References

Armstrong, C. (1994). Co-operative solutions in a transboundary fishery: The Russian–Norwegian co-management of the Arcto-Norwegian cod stock. *Marine Resource Economics, 9,* 329–351.

Armstrong, C., & Flaaten, O. (1991). The optimal management of a transboundary renewable resource: The Arcto-Norwegian cod stock. In R. Arnason & T. Bjørndal (Eds.), *Essays on the economics of migratory fish stocks* (pp. 137–152). New York: Springer.

Bacharach, M. (1976). *Economics and the theory of games*. London: MacMillan Press.

Barrett, S. (2003). *Environment and statecraft: The strategy of environmental treaty making*. Oxford: Oxford University Press.

Bjørndal, T., & Lindroos, M. (2004). International management of North Sea herring. *Environmental & Resource Economics, 29,* 83–96.

Bjørndal, T., & Lindroos, M. (2012). Co-operative and non-cooperative management of the Northeast Atlantic cod fishery. *Journal of Bioeconomics, 14,* 41–60.

Bjørndal, T., & Munro, G. (2012). *The economics and management of world fisheries*. Oxford: Oxford University Press.

Food and Agriculture Organization of the UN. (1992). *Marine fisheries and the law of the sea: A decade of change*. FAO Fisheries Circular No. 853. Rome: FAO.

Gordon, H. S. (1954). The economic theory of a common property resource. *Journal of Political Economy, 62,* 124–142.

Gravelle, H., & Rees, R. (1992). *Microeconomics*. Essex: Addison Wesley Longman Ltd.

Hnyilicza, E., & Pindyck, R. (1976). Pricing policies for a two-part exhaustible resource cartel: The case of OPEC. *European Economic Review, 8,* 139–154.

Miller, K., & Munro, G. (2004). Climate and cooperation: A new perspective on the management of shared fish stocks. *Marine Resource Economics, 19,* 367–393.

Miller, K., McDorman, T., McKelvey, R., Munro, G., & Tydemers, P. (2001). *The 1999 Pacific Salmon Agreement: A sustainable solution?* Canadian–American Public Policy, Occasional Papers No. 47. Orono: Canadian–American Center, University of Maine.

Munro, G. (1979). The optimal management of transboundary renewable resources. *Canadian Journal of Economics, 12,* 355–376.

Munro, G. (2013). Regional fisheries management organizations and the new member problem: From theory to policy. In A. Dinar & A. Rapoport (Eds.), *Analyzing global environmental global issues* (pp 105–128). London: Routledge.

Nash, J. (1953). Two-person cooperative games. *Econometrica, 21,* 128–140.

Owen, G. (2013). *Game theory* (4th ed.). Bingley: Emerald Group Publishing Ltd.

Treaty Between the Government of Canada and the Government of the United States of America Concerning Pacific Salmon. (2009). Retrieved from http://www.psc.org/pubs/treaty.pdf.

United Nations. (1982). *United Nations convention on the law of the sea.* UN Doc. A/Conf.62/122.

von Neumann, J., & Morgenstern, O. (1944). *Theory of games and economic behavior.* Princeton: Princeton University Press.

Chapter 5
Cooperative Games in Fisheries with More than Two Players

Abstract The cooperative game presents the first-best solution to fisheries management. It can be referred to as the *good-neighbours* or the *social optimal* solution and is often used for welfare comparisons. The chapter presents the characteristic function approach, also known as the *cake-sharing* model. It is the model concerned with the division of the net return from cooperation. International fisheries agreements (IFAs) are usually considered as inherently fragile. This chapter presents the theoretical foundation for stable IFA and the need to neglect nothing that may strengthen a given IFA, hereunder also side payments. It highlights the challenges when applying the characteristic function approach to renewable resource problems. The chapter concludes by presenting sharing rules and discussing stability issues both theoretically and in real-world cases.

5.1 An Introduction to Cooperative Games with More than Two Players: The Importance of Coalitions

The previous chapter, concerned solely with two-player cooperative games, makes the point that there is a significant difference between $n = 2$ and $n > 2$ cooperative games, with the former being much, much simpler than the latter. The simple $n = 2$ cooperative games, as discussed in Chap. 4, are undoubtedly useful in analysing the cooperative management of strictly transboundary fishery resources. They are, however, hopelessly inadequate for analysing the cooperative management of straddling fish stocks, which as Chap. 1 makes clear has been a major fishery management issue since the late 1980s. The typical Regional Fisheries Management Organization (RFMO) called for by the UN Fish Stocks Agreement (United Nations 1995) for the management of straddling stocks has a large number of members (players). To take but one example, the RFMO, the Northwest Atlantic Fisheries Organization (NAFO), has twelve contracting parties; a not exceptionally large RFMO.

When a potential cooperative game has $n > 2$, the issue of coalitions must be taken seriously. In Chap. 2, it was pointed out that the number of coalitions in a cooperative game is equal to 2^n. Thus, in a $n = 2$ cooperative game, there are four coalitions: the Grand Coalition (the two players playing cooperatively), two singleton coalitions

© Springer Nature Switzerland AG 2020
L. Grønbæk et al., *Game Theory and Fisheries Management*,
https://doi.org/10.1007/978-3-030-40112-2_5

and the empty coalition, Ø. Our interest lies solely in whether the two players do, or do not, cooperate. One need not be unduly interested in coalitions. Suppose, however, that $n = 3$, with the result that the number of possible coalitions is equal to 8. Denoting the players as P1, P2 and P3, one must be concerned, not only with all three playing cooperatively (Grand Coalition) and the singleton coalitions, but also with the possibility that P1 and P2 may form a sub-coalition and play competitively against P3, or that P2 and P3 may form a sub-coalition and play competitively against P1, and so on. If $n = 12$, as in the case of NAFO, the number of possible coalitions is equal to 4096! Once $n > 2$, there is no choice but to enter the realm of coalition game theory. For a review of coalition game theory, see Lindroos et al. (2007).

Bloch (2003) sets forth three basic questions concerning endogenous coalition formation:

(1) which coalitions will be formed?
(2) how will the coalitional value be divided among coalition members?
(3) how does the presence of other coalitions affect the incentive to cooperate?

Questions (1) and (3) concern the serious problem of stability of the Grand Coalition. In Chap. 4, it is assumed that a cooperative arrangement upon being achieved would prove to be binding through time. That assumption is no longer acceptable with $n > 2$, where forces that serve to undermine the binding nature of the cooperative arrangement abound.

Question (2) is concerned with "fair" allocation of the cooperative benefits, the *cake-sharing* problem.[1] This problem, which is a simple one when $n = 2$, is very complex when $n > 2$.

It is simply not feasible to attempt to address fully all three questions in one chapter. This chapter will focus primarily on the "fair" allocation problem. References will be made to the stability issue, but these references will be cursory in nature. The stability issue will be addressed in detail, in the chapter to follow.

As a first step towards addressing the complex *cake-sharing* problem when $n > 2$, we introduce the key concept of characteristic function games.

5.2 The Concept of Characteristic Function Games

The characteristic function games constitute a branch of the theory of cooperative games, in which it is assumed that transferable utility prevails throughout.[2] Critically, the characteristic function assigns a single number value to each possible coalition, with that value being equal to what in Chap. 4 had been termed the cooperative

---

[1]It is important to distinguish this problem from the *cake-cutting* problem. In the case of *cake-sharing*, the claims do not exceed the total worth of the coalition. In the case of *cake-cutting*, the claims do exceed to total worth and the problem turns into a bankruptcy problem instead.

[2]In Chap. 4, reference was made to side payments. Side payments and transferable utility are essentially one and the same. To say that transferable utility prevails throughout is to say that are no constraints on the use of side payments.

surplus, namely, the payoff from the coalition in excess of the sum of payoffs that would be received by the coalition players, if they played competitively. By definition, the characteristic function values of the empty coalition and the singleton coalitions all equal zero.

If $n = 2$, there would be but one coalition to which the characteristic function would assign a positive value, the Grand Coalition. This is not the case when $n > 2$. Return to our $n = 3$ example. It is then very possible that, as well as the Grand Coalition having a positive value, the characteristic function would assign a positive value to say the P1, P2 sub-coalition. This will affect the bargaining over the sharing of the Grand Coalition value. P1 and P2 would never accept a sharing arrangement, which would see their combined share of the Grand Coalition value being less than that of the value of the P1, P2 sub-coalition.

The introduction of coalition games and the *characteristic function games* (often referred to as *c-games*) in the fisheries literature came about as a consequence of international fisheries agreements (IFAs) in the form of RFMOs achieving prominence. That introduction must be attributed to Kaitala and Lindroos (1998), who are to be seen as pioneers in the application of c-games to fisheries economics.

5.2.1 The Theory Behind the Characteristic Function

For the definition of a coalitional game with transferable utility, it is assumed that there exists a finite set of n-players. Define the Grand Coalition as $N = \{1, 2, \ldots, n\}$, corresponding to the coalition formed by all n-players. Define $S = \{1, 2, \ldots, s\}$, such that $S \subseteq N$. S is then any of the 2^n sub-coalition of N including the empty coalition, $\emptyset$, and the Grand Coalition. S consists of s players, with $0 \leq s \leq n$.

Definition 5.1 An n-person cooperative game in the characteristic function form is an ordered pair $G = \langle N, v \rangle$ where $N = \{1, 2, \ldots, n\}$ is a finite set with n elements, and where v denotes the characteristic function. The characteristic function for a coalition S, $v(S)$, is defined as a function that associates with every non-empty subset S of N, a real number $v(S)$.

The characteristic function for a coalition S in absolute terms is often referred to $\bar{v}(S)$. That number is the coalition worth, which it will be recalled, corresponds to the total payoff of the coalition minus the sum of the payoffs of its members (players) under full non-cooperation (Mesterton-Gibbons 2000).

For the sake of convenience, the characteristic function is normalized with respect to the Grand Coalition. The normalized characteristic function is thus defined as $v(S) = \frac{\bar{v}(S)}{\bar{v}(N)}$. The complete characteristic function is determined by assigning a value to each of the 2^n sub-coalitions, with the empty coalition, $v(\emptyset) = 0$. A coalitional game with transferable payoff or utility is usually represented as $\langle N, v \rangle$.

In defining the characteristic function, the formation of a coalition S, and the corresponding payoff to S, consideration is not directly taken into account of the effects on this corresponding payoff of the n–s players outside the coalition S. In point of fact, the determination of the payoff to coalition S is influenced by the nature of the game between S and the other coalitions.[3]

The most widely accepted assumption is that the game between S and the other coalitions is a γ-type of game, a Nash non-cooperative game, in which the players adopt their mutually best responses (Chander and Tulkens 1997). Chander (2007) claims that the γ-characteristic function is the rational approach. Finus (2003) argued the γ-characteristic can impose a weak punishment on deviators compared to other characteristic function (e.g. the α-characteristic function, which is a maxi-min strategy where players outside the coalition adopts the strategy implying worst possible outcome for the coalition).

Definition 5.2 The characteristic function $v \in G$ is *superadditive* if, for all $S, T \subseteq N$ with $S \cap T = \emptyset$, $v(S \cup T) \geq v(S) + v(T)$.

Definition 5.2 implies that a union of two coalitions with no overlap of members will never diminish their joint benefits. When a game is superadditive, it implies there is always incentives to cooperate since cooperation increases (or equals) the benefits.

There is one caveat, however. As will become apparent in Chap. 6, coalition formation in fishery games may generate externalities. That is to say, the formation of a coalition may have an external effect on the payoffs of the players outside the coalition. If such externalities are generated by coalition formation, superadditivity may fail to hold. All of this must wait until Chap. 6.

5.2.2 The Characteristic Function Applied to a Simple Fishery Model

To gain a better understanding of how the theory of the characteristic function can be applied to fishery models, we develop an example based upon the steady-state bioeconomic model introduced in Chap. 3. As in Chap. 3, we apply the standard Gordon–Schaefer model (Gordon 1954; Schaefer 1954), assuming that n-players exploit the stock:

$$F(X) = rX\left(1 - \frac{X}{k}\right) \tag{5.1}$$

$$H_i(E_i, X) = qE_iX \tag{5.2}$$

---

[3] The singletons are also referred to as coalitions.

$$F(X) = \sum_{i=1}^{n} H_i(E_i, X), \tag{5.3}$$

where X represents the biomass of fish stock and $F(X)$ is the natural growth function, assumed to follow the logistic growth as described in Eq. (5.1). $H_i(E_i, X_t)$ is the harvest by player i employing effort E_i to the fishery. r and k are biological parameters defined as the intrinsic growth and the carrying capacity, respectively. The harvest for player i is assumed to be linear in effort for player i, E_i, and the biomass. q is the catchability coefficient, which is assumed identical for all n-players.[4]

From Eqs. (5.1)–(5.3), the relation between the steady-state stock level and the fishing efforts can be obtained:

$$X = \frac{k}{r}\left(r - q\sum_{i=1}^{n} E_i\right). \tag{5.4}$$

Now assume that the price of harvested, p, is constant and common to all n-players. Next, we assume that the marginal cost effort for any given player i, c_i, is constant. We shall further assume, however, that the marginal cost of effort is asymmetric among players; there are low-cost players, high-cost players and those in between. Since each player has constant marginal cost of effort, a coalition will always employ the most efficient technology (lowest c_i) available in the coalition.[5]

With this information, the optimization problem for any of the possible 2^n sub-coalition of N can be defined as follows:

$$\max_{E_S} \pi(S) = pH_S(E_S, X) - c_i E_S,$$

$$\text{s.t. } X = \frac{k}{r}\left(r - q\sum_{i=1}^{z} E_i\right), \tag{5.5}$$

where z is defined as the number of coalitions formed by the players, where each player belongs to exactly one coalition. Being a singleton is also considered as a coalition.[6] More on this will come in Chap. 6. From Chap. 3, we know that the optimal effort for the individual players, which is now interpreted to be the coalitions, are defined as $E_i = \frac{zr}{(z+1)q}(1 - b_i) - \sum_{j \neq i} \frac{r}{(z+1)q}(1 - b_j)$, where b_j is the inverse efficiency parameter defined as $b_j = \frac{c_j}{pqk}$.

---

[4]The value of q depends on the units in which E is measured.

[5]In case the marginal cost were not constant, and several players in a coalition are active, then a coalition would always apply the effort with the lowest marginal cost and the switching point(s) would be according to the equi-marginal principle, e.g. where the marginal costs of effort for the players in the coalition equalize.

[6]If, for example, coalition $S = N$ then $z = 1$. If $S \subseteq N$ then z corresponds to the number of coalitions, which will be n if all players remain singletons.

Based on the optimal effort levels, we can define the payoff function $\pi(S)$, which maps to each coalition S it's payoff. The characteristic function for coalition S is defined as $\bar{v}(S) = \pi(S) - \sum_{i \in S} \pi(\{i\})$. The normalized characteristic function is then defined as

$$v(S) = \frac{\bar{v}(S)}{\bar{v}(N)} = \frac{\pi(S) - \sum_{i \in S} \pi(\{i\})}{\pi(N) - \sum_{i=1}^{N} \pi(\{i\})} \tag{5.6}$$

and expresses the coalition worth relatively to the coalition of the Grand Coalition. Hence by applying a simple bioeconomic model with a limited number of players, a coalition game can be formed and the characteristic function is applied to describe the normalized coalition value. The coalition worth is thus simply to be understood as the excess profits of the coalition compared to non-cooperation. For easier interpretation (interpretation as a ratio), this is normalized to the worth of the Grand Coalition.

5.2.3 A Numerical Example

Assume now a fishery that is exploited by three players, P1, P2 and P3, which are asymmetric in their effort cost parameters $c_1 < c_2 < c_3$, but which are otherwise identical. From the bioeconomic model presented in Sect. 5.2.2 and in Chap. 3, we can, if we know the parameter values, determine the profits arising from different coalition scenarios. Assume that we have the following parameter values: $r = 1$, $k = 100$, $q = 0.1$, $p = 1$, and $c_1 = 1$, $c_2 = 2$, and $c_3 = 3$.

By applying the model presented in Sect. 5.2.2, we can calculate the effort, the stock size and hence the profit for each coalition. Based on the profits, the characteristic function and normalized characteristic function can be derived.

Table 5.1 summarizes the effort, the profits, the characteristic function and the normalized characteristic function.

In the example, we have no extra coalition worth among the singletons; hence, the normalized characteristic function is zero for the singletons. This occurs by default, given the definition of the characteristic function.

In the two-player sub-coalitions, the extra coalition worth varies between 7 and 23% of the extra coalition worth arising from the Grand Coalition, where all fishermen cooperate in conserving the fish stock.

5.3 The Allocation Problem, Sharing Rules and Side Payments

With the discussion of characteristic function games complete, we are now in a position to address Bloch's question 2, on how the coalition value will be divided

Table 5.1 Profits to different coalitions in a fishery

Coalition, S	Effort, E_S	Profits, $\pi(S)$	Characteristic functions, $\bar{v}(S) = \pi(S) - \sum_{i=1}^{S} \pi(\{i\})$	Normalized characteristic function, $v(S) = \frac{\bar{v}(S)}{\bar{v}(N)}$
$\emptyset$	–	–	–	–
$\{1\}$	3	9	$9 - 9 = 0$	$\frac{0}{6.25} = 0$
$\{2\}$	2	4	$4 - 4 = 0$	$\frac{0}{6.25} = 0$
$\{3\}$	1	1	$1 - 1 = 0$	$\frac{0}{6.25} = 0$
$\{1, 2\}$	3.67	13.44	$13.44 - (9 + 4) = 0.44$	$\frac{0.44}{6.25} = 0.07$
$\{1, 3\}$	3.33	11.11	$11.11 - (9 + 1) = 1.11$	$\frac{1.11}{6.25} = 0.18$
$\{2, 3\}$	2.33	5.44	$5.44 - (4 + 1) = 1.44$	$\frac{1.44}{6.25} = 0.23$
$\{1, 2, 3\}$	4.5	20.25	$20.25 - (9 + 4 + 1) = 6.25$	$\frac{6.25}{6.25} = 1$

Note $1, 2, 3 \doteq P1, P2, P3$, numbers are subject to rounding

among coalition members. Recall to begin that is being assumed that the coalitional game is one with transferable payoff or utility, $\langle N, v \rangle$, i.e. side payments can occur.

The only discussion that we have had to this point of the division of the returns from cooperation has been in Chap. 4 where the Nash bargaining solution is discussed. It calls for the cooperative surplus to be divided equally between/among the players. This is perfectly reasonable when $n = 2$ but may be entirely unreasonable when $n > 2$. Other approaches are needed. This section brings up the allocation problem inside the coalition, by defining sharing rules and discussing side payments. The importance of finding a "fair" allocation of the benefits from cooperation is that, without the assurance of a "fair" allocation, the effort to achieve stable cooperation may be stillborn.

By applying the information from the characteristic function game, it is possible to find a distribution or allocation of the aggregate worth from cooperation among players in any sub-coalition S. It is a payoff vector of real numbers, and it is often referred to as a sharing imputation or sharing rule. This application is, of course, closely tied to the assumption of transferable utility. For instance, the Nash bargaining solution, the nucleolus and the Shapley value are commonly used fair sharing rules, which are based on different fairness concepts. The Nash bargaining solution with equal weights yields an "egalitarian", or equal sharing, imputation of the gains. The nucleolus maximizes the minimum gains to any possible coalition, that is, the gains of the "least satisfied coalition", and the Shapley value divides the gains according to each player's average contribution to the coalitional worth.

Before turning the attention to the different sharing rules, the formal properties of the payoff vector or imputation of the characteristic function are defined.

Definition 5.3 Let $v \in G$. An imputation of v is a vector $x \in \mathbb{R}^n$ such that
$\sum_{i \in N} x_i = v(N)$ (group rationality);
$x_i \geq v(\{i\})$, $\forall i \in N$ (individual rationality).
The set of all imputations is denoted by X.

Definition 5.3 defines the framework for the sharing imputation. For the allocation to be individual rational (*Nash's individual rationality constraint*), it must allocate at least what could have been obtained by not joining the coalition. In the definition of the characteristic function $v(\{i\}) = 0$ and hence $x_i \geq 0$, $\forall i = 1, 2, \ldots, n$. For the sharing rule to be group rational all benefits in the coalition must be shared among the coalition members and hence $x_1 + x_2 + \cdots + x_n = 1$.

The payoff vector is thus a n-dimensional vector, $x = (x_1, x_2, \ldots, x_n)$, where the ith element of the vector, x_i, defines the share of the benefits allocated to player i.

5.3.1 Nash Bargaining Solution

We commence with the Nash bargaining solution as defined in Chap. 4 in Sect. 4.1.1, for a two-player cooperative game. We now at this point just require assurance that the bargaining solution holds when the number of players exceeds 2. The solution concept does, in fact, hold. It does not face major changes when extended to a game with more than two players. Hence, the Nash bargaining solution is the vector, $u \in \mathbb{R}^n$, that maximizes the Nash product $\max_{u_i} N(u) = \prod_{i=1}^{n} (u_i - \overline{u_i})$, s.t. $u_i > \overline{u_{i_{u_i}}} N$, $\forall i = 1, 2, \ldots, n$. Where u_i is the payoff to player i from the sharing of benefits from the Grand Coalition and $\overline{u_i}$ is the Threat Point or disagreement point that will occur if no agreement is formed. From this it follows that the Nash fair allocation rule extends to all cases in which the game has more than two players. The benefits from cooperation are to be allocated on an equal basis.

5.3.2 The Shapley Value

The Shapley value is a sharing rule based on the original contribution of Shapley (1953). The Shapley value is the first value function for a n-player c-game to be defined by axioms. The Shapley value assigns a value to each player which depends on his potential to change the value of the coalition by joining or leaving it. Said differently, the Shapley value can be interpreted as each member's expected marginal contribution to any coalition S. The Shapley value is an imputation that satisfies three axioms (1) it is additive, (2) it is symmetric and (3) it is resistant to a dummy player, where a dummy player is defined as a player that does not contribute to the coalition worth by joining the coalition. For a description of these three axioms please refer to the appendix. The special feature of the Shapley value is that it is defined with reference to other games, and not a single game in isolation since it is determined by the player's marginal contribution to possible coalitions he/she can be a member of. The Shapley value is a function that assigns a unique feasible payoff profile to each $\langle N, v \rangle$. The Shapley value for player i is defined as follows:

$$\psi_i(\langle N, v \rangle) = \frac{1}{N!} \sum_{S \in K, i \in S} (|N| - |S|)!(|S| - 1)!(v(S) - v(S \backslash \{i\}), \qquad (5.7)$$

where K is the set of all the possible coalitions, $|N|$ is the number of players in the game, $|S|$ is the number of players in coalition S, and $S \backslash \{i\}$ is coalition S without player i.

The Shapley value considers all possible orders, in which the Grand Coalition could actually form. By considering these orders and assuming that they have an equal likelihood of occurring, the marginal contribution of the individual players is calculated. The big advantage of the Shapley value is that it is easy to calculate, and because it has in its foundation the marginal contribution of each player, it is often considered a fair sharing rule.

A Numerical Example of the Shapley Value

The following example reuses the numbers from the numerical example introduced in Sect. 5.2.3. From the definition of the characteristic function we know that: $\bar{v}(P1) = \bar{v}(P2) = \bar{v}(P3) = 0$. In addition to the empty coalition, $\emptyset$, and the singleton coalitions, we have the following four possible coalitions:

(a) $\{P1, P2\}$,
(b) $\{P1, P3\}$,
(c) $\{P2, P3\}$ and
(d) $\{P1, P2, P3\}$.

Doing the usual normalization, the Nash bargaining solution as presented in Sect. 5.3.1 and thoroughly defined in Chap. 4 would give us the following, if a stable Grand Coalition were to be established:

$$x_1^{NB} = x_2^{NB} = x_3^{NB} = \frac{1}{3}.$$

The players have identical disagreement points, equal to zero, if no agreement is formed. To continue the example, we recall from Table 5.1 that we have the following: $\bar{v}(P1, P2) = 0.44$, while $\bar{v}(P1, P3) = 1.11$ and $\bar{v}(P2, P3) = 1.44$. The characteristic function for the Grand Coalition is $\bar{v}(P1, P2, P3) = 6.25$. The Shapley value payoffs imply that, if the Grand Coalition is established, the additional payoffs to P1 from cooperation in absolute terms will be 1.86. This is determined using Eq. (5.7). For player 1, it looks like follows:

$$\psi_1 = \frac{1}{3!}\Big\{(3-3)!(3-1)!(v(P1, P2, P3) - v(P2, P3))$$

$$+ (3-2)!(2-1)!(v(P1, P2) - v(P2)) + (3-2)!(2-1)!(v(P1, P3) - v(P3))$$

$$+ (3-1)!(1-1)!(v(P1) - v(\varnothing))\Big\}$$

$$= \frac{1}{6}\Big\{1 \cdot 2 \cdot \left(\frac{6.25}{6.25} - \frac{1.44}{6.25}\right) + 1 \cdot 1 \cdot \left(\frac{0.44}{6.25} - 0\right) + 1 \cdot 1 \cdot \left(\frac{1.11}{6.25} - 0\right) + 2 \cdot 1(0-0)\Big\} \approx 0.298.$$

The share (29.8%) refers to what P1 achieves of the total additional benefits (the characteristic function $\bar{v}(S)$) from the Grand Coalition. Similar it can be calculated that $\psi_2 = 32.4\%$ and $\psi_3 = 37.8\%$. In absolute terms, this corresponds to 29.8% of 6.25 corresponding to $0.298 \cdot 6.25 = 1.86$. The absolute additional payoff for P2 and P3, respectively, will be 2.03 and 2.36. At a first glance, it may seem counter-intuitive that $\psi_1 < \psi_2 < \psi_3$, but one has to keep in mind that this is the allocation of the additional benefits that can be created from the Grand Coalition. It becomes eminently reasonable when the total payoff the players is calculated, hence P1 receives $\pi(\{1\}) + \psi_1 \cdot \bar{v}(\{1, 2, 3\}) = 9 + 1.86 = 10.86$. If P1's total payoff from the Shapley value is compared to the total payoff from forming a Grand Coalition, $\pi(\{1, 2, 3\}) = 20.25$, it becomes obvious that P1 receives more than half of these benefits, while P2 and P3 receive 6.03 and 3.36, respectively. Hence, the intuition that P1 has the lowest cost and therefore should receive the largest share of the benefits holds true.

The numerical example is illustrated with the numbers derived from a steady-state model in a previous section of this chapter. An alternative version is to apply a dynamic Clark model as presented in Chaps. 3 and 4. For inspiration please refer to the model in Kaitala and Lindroos (1998) with the management of straddling and highly migratory stocks. They suppose that we have three players, P1, P2, and P3, where player 1 is a coastal nation and player 2 and 3 are both distant water fishing states (DWFS), where the DWFS have a particular interest in developing an effective programme for conservation and management of the stock. The players are identical in all other respects, except for fishing efforts costs. Denoting unit effort costs as c, we have: $c_1 < c_2 < c_3$. Kaitala and Lindroos (1998) demonstrate that in this specific example the Shapley value payoffs are as follows: $\psi_1 = \psi_2 = \frac{1}{3} + \frac{v(P1,P2)}{6}$ and $\psi_3 = \frac{1}{3} - \frac{v(P1,P2)}{3}$. This reflects the fact that the bargaining power of P1 and P2 exceeds that of P3. If it is not possible to establish the Grand Coalition, P3 ends up playing as a singleton. P1 and P2, on the other hand, have the fallback of establishing a sub-coalition and playing competitively against P3. This is a slightly different conclusion from the presented numerical example and is due to the special elements of this linear dynamic model where $\bar{v}(P1, P3) = \bar{v}(P2, P3) = 0$.

For the sake of completeness, we also present an alternative approach towards calculating the Shapley value, which strongly leans on the intuition behind the marginal contribution of the individual player to the coalition depending on the order the player joins the coalition. Imagine we would like to calculate the marginal contribution of P1 to any possible order of formation of the Grand Coalition. We first need to set up the possible orders of which the players can join the Grand Coalition and then calculate the marginal contribution of P1 in each of the possible orders of formation of the

Table 5.2 The marginal contribution of player 1

Order	Marginal contribution, P1	Actual marginal contribution
1, 2, 3	$v(\{1\}) - v(\{\emptyset\})$	$0 - 0 = 0$
1, 3, 2	$v(\{1\}) - v(\{\emptyset\})$	$0 - 0 = 0$
2, 1, 3	$v(\{1, 2\}) - v(\{1\})$	$0.07 - 0 = 0.07$
2, 3, 1	$v(\{1, 2, 3\}) - v(\{2, 3\})$	$1 - 0.23 = 0.77$
3, 1, 2	$v(\{1, 3\}) - v(\{1\})$	$0.18 - 0 = 0.18$
3, 2, 1	$v(\{1, 2, 3\}) - v(\{2, 3\})$	$1 - 0.23 = 0.77$
	Sum of above	1.79
	Shapley value, ψ_1	$\frac{1}{6} \cdot 1.79 = 29.8\%$

Grand Coalition. For example, in (1, 2, 3) P1 arrives as the first player, P2 as second and P3 as third. P1's marginal contribution is thus the addition to the existing empty coalition, e.g. $v(\{1\}) - v(\{\emptyset\})$. In the Grand Coalition orders (3, 2, 1) P3 arrives first, P2 second and P1 third, hence P1's marginal contribution is $v(\{1, 2, 3\}) - v(\{2, 3\})$. This process is continued for all possible orders of forming the Grand Coalition and the Shapley value is calculated. All these possible ways of ordering the Grand Coalition are summarized in Table 5.2 together with the payoffs.

5.3.3 The Nucleolus

The nucleolus is a single point in the core with the property that it minimizes the maximum dissatisfaction. It is the lexicographic centre of the core and is the imputation that maximizes the minimum gain to any possible coalition. Calculating the nucleolus requires defining the reasonable set, the excess function, and the core and then determining the nucleolus. In this section, these elements are briefly described, the appendix allows for more details on the elements.

Reasonable Set
The *reasonable set* is a set of imputations usually determined as a fair distribution of benefits. It is determined by the set of imputations X in Definition 5.3 with the constraint that no player receives more than what the player contributes to the coalition.

$$x_i \leq \max_{T \in \Pi^i} \{v(T) - v(T - \{i\})\}, \text{ where } \Pi^i = \{S | i \in S \wedge S \subseteq N\} \text{ and } N = \{1, 2, \ldots, n\}.$$

Thus, a fair allocation which ensures that no one receives more than the maximum they contribute to the Grand Coalition.

Excess

Based on the reasonable set, *the excess* is defined as the difference between the fraction of the benefits of cooperation that S can obtain for itself (from the normalized c-function) and the fraction of the benefits of cooperation that x allocates to S. The excess is thus

$$e(S, x) = v(S) - \sum_{i \in S} x_i.$$

The Core

The excess is useful for defining the core. The modern concept of *the core* was introduced by Gillies (1953). The core is a set of feasible imputations which cannot be improved by any coalition. The core is defined by the excess being negative and is added as a condition to the reasonable set. It is the imputations that are preferred to any other coalitions, e.g. ensures that every coalition receives less benefits than the Grand Coalition, $C^+(0) = \{x \in X | e(S, x) \leq 0, \forall S\}$. The core may be empty. If the nucleolus exists, it is always in the core. The same is not true for the Shapley value.

The Nucleolus

If the core is non-empty, the boundaries of the core are shrunk until they collapse (shrinking further would imply the empty set). This is the definition of the *least rational ε-core*, often this is also referred to simply as the least core. If the least rational ε-core results in a single point, that point is identical to *the nucleolus*. If the set does not correspond to a single point, the nucleolus if found by the concept of minimizing maximum dissatisfaction. This requires a definition of the set of coalitions whose excess can be reduced below the least rational ε-core and finding the minimax in this set. This process continues until it collapses to a single point, which is referred to as the nucleolus. The intuition is that the boundaries of the core are shrunk with the same rate for all players (corresponds to the ε), then it defines the minimum set of imputations (sharing allocations). The nucleolus is then determined among this set, which is done by minimizing the maximum dissatisfaction. For a technical description of the process of finding the nucleolus if the least rational ε-core is larger than a single point, please refer to the appendix. In Box 5.1, the reader finds an overview of the steps towards finding the nucleolus.

Box 5.1
Overview of the steps for finding the nucleolus

Step	Name	Meaning
1	Reasonable set	The set is defined by: – players receive no more than their actual contribution – individual rationality – group rationality
2	Excess function	The function defines the excess from a sharing imputation to any possible coalition
3	Core	The core is defined by a negative excess
4	Least core	By shrinking boundaries of the core with the same rate, stopping at the last point before it collapses if shrunk further
5	Nucleolus	The single point that minimizes the maximum dissatisfaction of complaint that any coalition could have against the imputation

5.3.4 The Satisfactory Nucleolus

In many IFA games, externalities from coalition formation will be present. To understand this, let us remind ourselves about the numerical example presented in Sect. 5.2.3. We presented the effort and profits for 2^n possible coalitions, but in order to derive these profits, we also needed to derive the effort for the players outside the coalitions; otherwise the stock would be unknown to us. This is referred to as the whole partition, where each player is member of exactly one coalition. Chapter 6 will go more into details on this. Table 5.3 presents the profits and effort of the possible partitions in the numerical example from Sect. 5.2.3.

Table 5.3 elaborates on the simple numerical example presented in Sect. 5.2.3. From the table, it becomes obvious that there are differences in the profit levels for the players playing on their own depending on whether the other players form a coalition or not. For example, for P3 the profit is 1 if the other players are also

Table 5.3 Numerical example with three players, effort and profits

Partition	Effort, E_S	Profits, $\pi(S)$
{1}, {2}, {3}	3, 2, 1	9, 4, 1
{1, 2}, {3}	3.67, 1.67	13.44, 2.78
{1, 3}, {2}	3.33, 2.33	11.11, 5.44
{2, 3}, {1}	2.33, 3.33	5.44, 11.11
{1, 2, 3}	4.5	20.25

acting as a singleton while it is 2.78 if the other players form a two-player coalition. This phenomenon is due to externalities from coalition formation and is explained to be the case where a player outside the decision-making is affected by coalition formation of others. In the illustrated case, it is a positive externality, where the outside player gains from others forming a coalition. In most fisheries games externalities are present.[7] The consequence is that playing as a singleton is more beneficial if the other players are forming a coalition. This is commonly referred to as free riding. The Shapley and nucleolus sharing rules presented in this chapter does not consider the externalities from the game and that free riding becomes more attractive when others form a coalition. The consequence is that sometimes these sharing rules allocate less benefits to a player than what he/she could have obtained by free riding.

The satisfactory nucleolus includes the consequences of fishery games facing positive externalities. It implies only searching for sharing imputations among those satisfying the individual satisfaction. Hence, this corresponds to changing the framework for sharing imputations to search among, but technically to be found as the nucleolus.

Definition 5.4 Let $v \in G$. An imputation of v is a vector $x \in \mathbb{R}^n$ such that
$x_i \geq \frac{v(\{i_{\text{FreeRider}}\})}{v(N)}$, $\forall i \in N$ (individual satisfaction).

The set of all imputations is denoted by X.

An example of the application of the satisfactory nucleolus will be presented in the case study in Sect. 5.4.

5.4 Case Study: Example of a Model Applied to the Baltic Cod Fishery

Kronbak and Lindroos (2007) present the case of the Baltic cod fishery. They apply an age-structured biological model for the cod fishery, assuming that three countries simultaneously exploit the resource. The bioeconomic model in their paper is slightly different from the simple standard Gordon–Schaefer model presented in previous sections. It does, however, not change the insights of the coalition model and the characteristic function. The objective function for the players is their net present values of instantaneous profit over 50 years with the limitation of assuming the same fishing strategy for all 50 years (an open-loop control). The countries aim to maximize their net present value of profits under the open-loop control, where countries' strategies

---

[7] As said, most games in fisheries face positive externalities; in fact, the authors have difficulties coming up with examples of fisheries games without externalities. It is not all games in other fields that faces positive externalities. As example, if you consider a carpool game (where potential congestion and pollution are ignores) then a third player would not face a positive externality from two other people carpooling.

Table 5.4 Characteristic function in the case of the Baltic cod fishery

Coalition	Strategy, *fishing mortality*	Characteristic function ($\bar{v}(S)$) (Dkr)	Normalized characteristic function ($v(S)$)	Normalized free-rider values
1	0.35	0	0	0.381
2	0.29	0	0	0.282
3	0.27	0	0	0.271
1, 2	0.457	2.76×10^9	0.143	
1, 3	0.457	2.57×10^9	0.133	
2, 3	0.407	1.20×10^9	0.062	
1, 2, 3	0.351	1.93×10^{10}	1	

Source Modified from Kronbak and Lindroos (2007)

are defined in terms of fishing mortality. The players are assumed to have quadratic cost functions but with different cost parameters. The difference in the cost parameters creates asymmetry among the players and due to the quadratic structure of the cost function the application of the equi-marginal principle is necessary. Table 5.4 presents the characteristic function and the normalized characteristic function obtained.

The Grand Coalition provides the economic efficient outcome (here defined as the maximal possible benefits). This has by default to be the case since the Grand Coalition can adopt any possible strategy and therefore also the strategy of a smaller coalition if this should prove beneficial. Hence, the Grand Coalition has the highest value for the characteristic function (the largest number in column 3). The characteristic function of any smaller coalition is between zero and the value of the Grand Coalition. Column 5 in Table 5.4 presents the normalized free-rider values of the players. The values are the normalized payoffs, relatively to the Grand Coalition, when a player acts as a singleton and the other two players form a two-player coalition.

From the numbers, it is possible to apply the sharing rules and derive different sharing imputations. In this case, the Shapley value is $(\psi_1, \psi_2, \psi_3) = (0.359; 0.323; 0.318)$, and the nucleolus is $(x_1, x_2, x_3) = (0.333; 0.333; 0.333)$ corresponding also in this case to the egalitarian solution. It becomes clear that the Shapley value to a large extent includes the bargaining power of the single players and therefore allocates a larger share to player 1 and a smaller to player 3 relatively. The nucleolus finds a lexicographic centre and, in this case, yields an equal sharing of the benefits.

Kronbak and Lindroos (2007) use a combination of a c-game and a non-cooperative coalition game to analyse the consequences of externalities to cooperative agreements. The authors reformulated the nucleolus sharing imputation, where the nucleolus is found by asking each coalition S about its dissatisfaction with the proposed sharing imputation and then by minimizing the maximum dissatisfaction from the sharing imputation. The reformulation by Kronbak and Lindroos (2007) included taking free-rider incentives into consideration, rather than the non-cooperative Nash solution. They called this adjusted approach the satisfactory nucleolus. The authors applied the alternative approach to the Baltic Sea cod fishery and demonstrated that this approach may stabilize an agreement, also in cases where the traditional sharing imputations such as the Shapley and the nucleolus cannot stabilize it.

From Table 5.4, it is possible to calculate the excess benefits. The benefits of the Grand Coalition correspond to 1 since the characteristic function is normalized (column 4 in the table). The normalized sum of benefits for the singletons when they are free riding on the Grand Coalition (numbers taken from the fifth column) is 0.934. This is derived as the sum of normalized free-rider benefits for player 1 (0.381), for player 2 (0.282) plus for player 3 (0.271). Hence, the sum of normalized benefits for the players if they individually act as free riders corresponds to 93.4% of the benefits of the Grand Coalition. It can, therefore, be concluded that the Grand Coalition can be stabilized (93.4 < 100%) if the appropriate sharing rule is applied. In the paper, the satisfactory nucleolus is calculated to $\left(x_1^s, x_2^s, x_3^s\right) = (0.403; 0.304; 0.293)$. The satisfactory nucleolus takes the Threat Points of non-cooperation into consideration and therefore also allocates a larger share to player 1 and smaller shares to player 2 and 3, respectively. Within games involving international fisheries agreements, the bioeconomic models often present games with positive externalities. Therefore, they are not necessarily free-rider stable, which leads to the application of an alternative sharing allocation and alternative approaches which are to be presented in more detail in Chap. 6.

5.5 Policy Implications

This chapter shows the complications of having more than two players in a game and still aiming towards an agreement with cooperative exploitation of a resource. As the chapter highlights, it is not straightforward to find a fair way of dividing the benefits. Furthermore, there is a high risk of having strong free riding incentives which easily makes agreements unstable. Hence when it is seen in the world of policy that IFAs are inherently fragile, it is important not to neglect the conditions that are underlying the stability of the agreement. To economists the crucial foundation of the agreements is the Nash's individual rationality constraint. This constraint says that a party will only join the IFA if he/she achieves at least the benefits that could have been obtained without the IFA. To economists this constraint may seem obvious since it is based on rationality. To policymakers it may seem less obvious since the common good may have higher priority. The rational behaviour does imply that side payments may be necessary to obtain the common good of an IFA. Side payments should be considered as a mean towards the goal. At a first glance, side payments may seem "immoral" to policymakers, but they should rather be regarded as a way to broaden the scope for bargaining, and thereby enhancing the prospects for long-term stability of the IFA. The tools provided in this chapter are different perspectives on how to determine the actual side payments. The chapter has been concerned with how to find a vector to describe how to share the benefits of IFAs, where players have an incentive to remain in the binding agreement. The rules, besides their mathematical expressions, have underlying intuitions which can be useful for policymakers, who should pay particular attention to three things: (1) Players remain sovereign and hence the risk of non-cooperation must be taken seriously. (2) Is there an excess payoff large enough from cooperation to make a binding agreement? And (3) Are side payments (or other types of transfers) needed to stabilize the agreements?

The chapter has not dealt with alternative threats or punishments if players do not comply with an agreement. Such elements could include a non-cooperative behaviour from the other players or governmental fines.

Applying the theory to bioeconomic models allows predictions concerning the likely implications of policy decisions. The theoretical setup sets the first-best cooperative outcome and defines the allocation to the single members. The model is built to predict the benefits and the stability of a joint IFA. In a bioeconomic context, this could, in some cases, be considered as the allocation of rights to harvest species. In other cases, the cooperative solution would require some states or regions to avoid harvesting for the common good, for instance, due to differences in cost structures. This is a well-known dilemma of the cooperative solution. The dilemma can be resolved through the use of side payments (Munro 2000).

Side payments are a transfer of benefits to players to induce them to join the agreement. The transfer of benefits between players is a necessary condition to reach the mutual best benefits from their joint agreement. The problem origins from the discussion of a potential versus an actual Pareto efficiency. The cooperative outcome presented in the characteristic function games is the potential Pareto efficient outcome. In order to obtain the actual Pareto improvement, side payments are the tool. Side payments can thus be seen as a means to stabilize an agreement.

Appendix

The Shapley Value
This part of the appendix describes the axiomatic approach of the Shapley value. Assumes a game with N-players and a value function, $v \langle N, v \rangle$.

Definition 5.5 Define player i as a dummy player if the marginal contribution of this player to any coalition is identical to its value when playing singleton, e.g. $v(S \cup \{i\}) - v(S) = v(\{i\})$.

Definition 5.6 Two players are interchangeable in the value function when the marginal contribution of the players is identical for every coalition or said differently the value is the same no matter if player i or player j is member of the coalition S, e.g. $v((S \setminus \{i\}) \cup \{j\}) = v(S)$.

The Shapley value is a function that assigns a unique feasible payoff profile to each player and satisfies the following three axioms:

(1) SYM (Symmetry)
     If player i and player j are interchangeable, then the Shapley value of the two players is identical, e.g. $\psi_i(\langle N, v\rangle) = \psi_j(\langle N, v\rangle)$.
(2) DUM (Dummy player)
     If player i is a dummy player, then the Shapley value of this player corresponds to its own value function, e.g. $\psi_i(\langle N, v\rangle) = v(\{i\})$.
(3) ADD (Additivity)
     This axiom imposes a link in the outcome of two games. It states the sum of the games is identical to a combined game. For any two games $\langle N, v\rangle$ and $\langle N, w\rangle$ we have $\psi_i(\langle N, v + w\rangle) = \psi_i(\langle N, v\rangle) + \psi_i(\langle N, w\rangle) \forall i \in N$, where $v + w$ is the game defined by $(v + w)(S) = v(S) + w(S)$ for every coalition S.

The Nucleolus

The nucleolus is a complex value to calculate. It is an iterative process and, as presented in this chapter, it may be helpful to divide the calculation in a series of five consecutive steps. This appendix will carefully address these steps. The discussion on the concept is followed by an illustrative example.

Step 1:   The reasonable set

To find the reasonable set we need to define the characteristics of the elements in the set. Often the notion of fairness is used with the argument that it is fair that players receive no more than their actual contribution, which is one of the conditions for an element to belong to the set. Furthermore, the reasonable set is founded by the necessary conditions for a stable solution as presented in Definition 4.1 for two players. More specifically the individual and collective rationality must hold.

> The *reasonable set* is determined by the set of imputations X in Definition 5.3 with the constraint that no player receives more than what the player contributes to the coalition. $x_i \leq \max_{T \in \Pi^i} \{v(T) - v(T - \{i\})\}$, where $\Pi^i = \{S | i \in S \wedge S \subseteq N\}$ and $N = \{1, 2, \ldots, n\}$.

Intuitively, the reasonable set is a set of fair allocations which ensures that no one receives no more than the maximum they contribute to the Grand Coalition.

Step 2:   The excess function

With a large number of sharing imputations, we need to be better informed about the characteristics of the different sharing imputations. It is assumed that agents are rational and maximize their benefits. Therefore, it is natural to be concerned about how much excess a coalition can obtain compared to individual players. This is what lays the foundation for defining the excess function. It is a function that defines the excess from a sharing imputation to any possible coalition. The excess function is defined for all elements of the reasonable set, as the difference between the fraction of the benefits of cooperation that S can obtain for itself (from the normalized c-function) and the fraction of the benefits of cooperation that x allocates to S.

This corresponds to the following definition of the excess function: $e(S, x) = v(S) - \sum_{i \in S} x_i$.

Step 3: The core

So far, we have a large set of sharing imputations, called the reasonable set, and a function, the excess function, to describe some of the characteristics of the elements in the set. To narrow the set a bit further down the concept of the core is applied. The core is a set of feasible imputations which cannot be improved by any coalition or, said differently, there is no positive excess in the elements contained in the core. The core is therefore defined by the sharing imputations in the reasonable set for which it is true that the excess function is negative (or zero). The core therefore contains imputations that are preferred to any other coalitions.

The definition of the core is $C^+(0) = \{x \in X | e(S, x) \leq 0, \forall S\}$.

The core may be empty, intuitively this happens if a coalition receives more benefits than the Grand Coalition.

Step 4: The least rational ε-core

To define the least rational ε-core requires a number of iterations. The idea is to define the set of coalitions whose excess can be reduced below the least rational ε-core and finding the minimax in this set. It may sound a bit complicated, below are the steps defined towards this.

First, define Σ^0 as the set of all coalitions that are neither the empty coalition nor the Grand Coalition, hence $\Sigma^0 = \{S | S \subset N, S \neq \emptyset\}$.

Second, suppose the boundaries of the core are moved inward with a rate ε. Call this the ε-core being defined by $C^+(\varepsilon) = \{x \in X | e(S, x) \leq \varepsilon, \forall S \in \Sigma^0\}$. This clearly resembles the definition of the core and in fact is the core for $\varepsilon = 0$.

Third, define the maximum excess function in Σ^0 by $\phi_0(x) = \max_{S \in \Sigma^0} e(S, x)$.

Fourth, use the ϕ_0-function to reduce the expression for the ε-core to $C^+(\varepsilon) = \{x \in X | \phi_0(x) \leq \varepsilon\}$. The boundaries of the rational core are moved inwards by reducing ε as much as possible without violating the condition $\phi_0(x) \leq \varepsilon$.

> This provides the formal description of the least rational ε-core to be $X^1 = C^+(\varepsilon_1)$, where $\varepsilon_1 = \min_{x \in X} \phi_0(x)$.

Fifth, if the least rational core contains a single point, this is the nucleolus. If the least rational core contains a larger set than a single point the set must be further reduced, the iteration starts with step 2 again. The next step is defining all sets of coalitions in the least rational core which have not reached the boundary ε_1.

Description of the iteration process:

1. Let Σ^1 be the set of all coalitions whose excess can be reduced below ε_1 by an imputation in the least rational core, X^1, $\Sigma^1 = \{S \in \Sigma^0 | e(S, x) < \varepsilon_1 \text{ for some } x \in X^1\}$.
2. Now define the maximum excess in least rational core among the set of imputations where the excess can be reduced below ε_1, hence in Σ^1. The maximum excess of coalitions in Σ^1 is $\phi_1(x) = \max_{S \in \Sigma^1} e(S, x)$.
3. The minimum of the maximum excess then becomes $\varepsilon_2 = \min_{x \in X^1} \phi_1(x)$.
4. The process continues with a set of all imputations at which the maximum excess of the coalitions in Σ^1 achieves its minimum, $X^2 = \{x \in X^1 | \phi_1(x) = \varepsilon_1\}$.
5. As may have become obvious to the reader, this is an iterative process that continues with keeping on constructing sub-sets of coalitions $\Sigma^0 \supset \Sigma^1 \supset \Sigma^2 \supset \cdots \Sigma^j$, excluding coalitions that cannot have excess reduced below $\varepsilon_1 > \varepsilon_2 > \cdots \varepsilon_j$ until $X = X^0 \supset X^1 \supset X^2 \supset \cdots X^j$ contains only a single imputation, where $\phi_0, \phi_1, \phi_2, \ldots, \phi_{j-1}$ are functions such that the domain of ϕ_j is X^j.

The intuition is shrinking the boundaries of the core with the same rate and stop at the last point, before it collapses if shrunk further.

Step 5: The nucleolus

Recall that the nucleolus is defined as the single point that minimizes the maximum dissatisfaction of complaint that any coalition could have against the imputation. We apply the least rational ε-core to get closer to the solution concept. If the least rational ε-core results in a single point, then it is identical to the nucleolus. If the least rational ε-core is a set larger than a single point, the nucleolus is found by the concept of minimizing maximum dissatisfaction. That is, to define the nucleolus for more than one ($j > 1$) recursions we need to apply the maxi-min excess ($\phi_j(x) = \max_{S \in \Sigma^j} e(S, x)$, $\Sigma^j = \{S \in \Sigma^{j-1} | e(S, x) < \varepsilon_j \text{ for some } x \in X^j\}$) and from this determine the least rational ε_j-core (($\varepsilon_j = \min_{x \in X^{j-1}} \phi_{j-1}(x)$) from which we define the set of all imputations at which the maximum excess of the coalitions in Σ^j achieves its minimum ($X^j = \{x \in X^{j-1} | \phi_{j-1}(x) = \varepsilon_{j-1}\}$). When X^j contains only a single point $X^* = \{x^*\}$, then x^* is the nucleolus.

Example of the Application of the Nucleolus

We will now demonstrate how to determine the nucleolus based on an example. The example is inspired by the Norwegian spring spawning herring fishery with three players (Lindroos and Kaitala 2000): player 1 (Norway with Russia), player 2

(Iceland with the Faroe Islands) and player 3 (the EU). Each player's strategy is to decide on the level of fishing mortality to apply in the fishery. From the bioeconomic model in Lindroos and Kaitala (2000), we have the following information (Table 5.5):

Step 1: The reasonable set

$$x_1 \leq \max\{v(N) - v(\{2, 3\}), v(\{1, 2\}) - v(\{2\}), v(\{1, 3\}) - v(\{3\}), v(\{1\}) - v(\emptyset)\}$$
$$= \max\{1 - 0.39, 0.34, 0.34, 0\} = 0.61,$$
$$x_2 \leq \max\{1 - 0.34, 0.34, 0.39, 0\} = 0.66,$$

$x_3 \leq \max\{1 - 0.34, 0.34, 0.39, 0\} = 0.66$, which corresponds to $x_1 + x_2 \geq 1 - 0.66 = 0.34$, as $x_3 = 1 - x_1 - x_2$. Moreover, from the definition of imputation (Definition 5.3), we have: $x_1 \geq 0$, $x_2 \geq 0$, and $x_3 \geq 0$. The latter condition can be rewritten as $x_1 + x_2 \leq 1$.

Based on the above calculations, the reasonable set is bounded by the six inequalities in the space (x_1, x_2) and can be illustrated as follows (Fig 5.1).

Step 2: The excess function

The excess function is determined as follows, remembering that $x_3 = 1 - x_1 - x_2$:

$$e(\{i\}, x) = -x_i, \text{ where } i = 1, 2, 3,$$
$$e(\{1, 2\}, x) = 0.34 - x_1 - x_2,$$
$$e(\{1, 3\}, x) = 0.34 - x_1 - x_3 = 0.34 - x_1 - 1 + x_1 + x_2 = -0.66 + x_2,$$
$$e(\{2, 3\}, x) = 0.39 - x_2 - x_3 = -0.61 + x_1.$$

Step 3: The core

The core is defined based on the excess function: $C^+(0) = \{x \in X | e(S, x) \leq 0, \forall S\}$. In our case, the core coincides with the reasonable set.

Table 5.5 The value of coalitions for the Norwegian spring spawning herring fishery	Coalition, S	Values of the coalitions	Free-rider value $(N{-}S)$	Normalized c-function $(v(S))$
	$\{1\}$	4878	–	0
	$\{2\}$	2313	–	0
	$\{3\}$	896	–	0
	$\{1, 2\}$	19,562	14,534 (player 3)	0.34
	$\{1, 3\}$	18,141	17,544 (player 2)	0.34
	$\{2, 3\}$	17,544	18,141 (player 1)	0.39
	$\{1, 2, 3\} = N$	44,494	50,219 Sum of above	1

Source Lindroos and Kaitala (2000), Table 3, and own calculations (normalized c-function)

Fig. 5.1 The reasonable set for the Norwegian spring spawning herring fishery

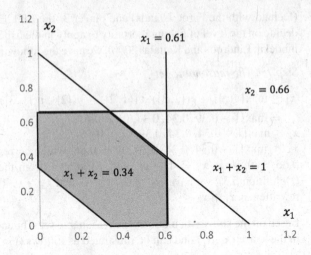

Step 4: The least rational ε-core.

The set of all coalitions which is not the empty coalition, nor the Grand Coalition is the following:

$$\Sigma^0 = \{\{1\}, \{2\}, \{3\}, \{1, 2\}, \{1, 3\}, \{2, 3\}\}.$$

The ε-core is defined as $C^+(\varepsilon) = \{x \in X | e(S, x) \le \varepsilon, \forall S \in \Sigma^0\}$ which in our case provides the following set of inequalities bounding the ε-core:

$$e(\{i\}, x) = -x_1 \le \varepsilon \Leftrightarrow -\varepsilon \le x_1 \tag{5.8}$$

$$-\varepsilon \le x_2 \tag{5.9}$$

$$-(1 - x_1 - x_2) \le \varepsilon \Leftrightarrow x_1 + x_2 \le \varepsilon + 1 \tag{5.10}$$

$$0.34 - \varepsilon \le x_1 + x_2, \tag{5.11}$$

$$x_2 \le \varepsilon + 0.66 \tag{5.12}$$

$$x_1 \le \varepsilon + 0.61. \tag{5.13}$$

Combining (5.8) and (5.13), we get

$$-\varepsilon \le x_1 \le \varepsilon + 0.61. \tag{5.14}$$

From this inequality, we have $-\varepsilon \le \varepsilon + 0.61$, and hence $\varepsilon \ge -0.305$.

Combining (5.9) and (5.12):

$$-\varepsilon \leq x_2 \leq \varepsilon + 0.66. \tag{5.15}$$

Using the lower and upper bound of x_2, we get an upper bound of $\varepsilon \geq -0.33$.
Combining (5.10) and (5.11):

$$0.34 - \varepsilon \leq x_1 + x_2 \leq \varepsilon + 1. \tag{5.16}$$

From this inequality, we obtain $\varepsilon \geq -0.33$.
Adding Eqs. (5.14) and (5.15):

$$-2\varepsilon \leq x_1 + x_2 \leq 2\varepsilon + 1.27. \tag{5.17}$$

Combining (5.16) and (5.10) yields two inequalities, $-2\varepsilon \leq \varepsilon + 1$ and $0.34 - \varepsilon \leq 2\varepsilon + 1.27$, whose solutions are $\varepsilon \geq -\frac{1}{3}$ and $\varepsilon \geq -0.31$, respectively.

Considering all cases above, the upper bound in the least rational core comes from inequality (5.14): $\varepsilon_1 = -0.305$. Inserting this bound into inequality (5.14) provides $x_1 = 0.305$. Hence, by using inequalities above the least rational core can be derived as $X^1 = \{x \in X | x_1 = 0.305, 0.34 \leq x_2 \leq 0.355, 0.34 \leq x_3 \leq 0.355\}$. The bounds for x_2 are obtained by inserting $\varepsilon = -0.305$ in inequality (5.15). As $x_3 = 1 - x_1 - x_2$, the lower bound of x_3 is obtained as one minus the sum of the upper bounds of x_1 and x_2 $(1 - 0.305 - 0.355 = 0.34)$. By the same token, the upper bound is one minus the sum of the lower bounds of x_1 and x_2 $(1 - 0.305 - 0.34 = 0.355)$. The least rational core is illustrated as the small thick black line segment in Fig. 5.2.

Since the least rational core contains more than a single point, the iterative process starts.

Fig. 5.2 The least rational core in the reasonable set

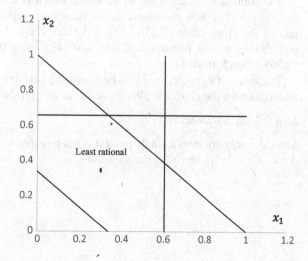

Fig. 5.3 The maximum
excess of the coalition. *Note*
Exp. 1 is $-x_2$, *Exp. 2 is*
$-0.695 + x_2$, *Exp. 3 is*
$0.035 - x_2$, *Exp. 4 is*
$-0.66 + x_2$, *and Exp. 5 is*
the new minimum excess

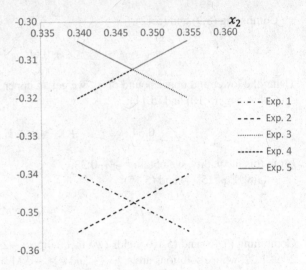

First, we define $\Sigma^1 = \{\{2\}, \{3\}, \{1, 2\}, \{1, 3\}\}$ which is the set of all coalitions whose excess can be reduced below $\varepsilon_1 = -0.305$, and where $x_1 = 0.305$.

Second, we derive the maximum excess of coalitions in Σ^1, that is $\phi_1(x) = \max_{S \in \Sigma^1} e(S, x)$. In our case, it corresponds to

$$[\phi_1(x_2)] = \max_{S \in \Sigma^1}(-x_2, -0.695 + x_2, 0.035 - x_2, -0.66 + x_2).$$

One way to determine the maximum of $\phi_1(x_2)$ is by drawing the four expressions it contains taking into consideration that $0.34 \leq x_2 \leq 0.355$. This is illustrated in Fig. 5.3.

The solid line indicates the ϕ_1 function, which is a combination of two of the expressions. The new minimum excess occurs when the two lines in the upward part of the graph cross ($0.035 - x_2 = -0.66 + x_2$). At this point $x_2 = 0.3475$ and thereby $\varepsilon_2 = \min \phi_1 = 0.035 - 0.3475 = -0.3125$. From this $X^2 = \{0.305, 0.3475, 0.3475\}$.

The value of ε_2 could have also been obtained using inequalities (5.8)–(5.13) as undertaken for the case of ε_1. We leave this as an exercise for the interested reader.

Step 5: The nucleolus

Since X^2 only contains a single point it is not possible to further minimize and thus $X^2 = X^* = \{x^*\}$, then x^* is the nucleolus.

References

Bloch, F. (2003). Non-cooperative models of coalition formation in games with spillovers. In C. Carraro (Ed.), *The endogenous formation of economic coalitions* (pp. 35–79). Cheltenham: Edward Elgar.

Chander, P. (2007). The gamma-core and coalition formation. *International Journal of Game Theory, 35,* 539–556.

Chander, P., & Tulkens, H. (1997). The core of an economy with multilateral environmental externalities. *International Journal of Game Theory, 26,* 379–401.

Finus, M. (2003). Stability and design of international environmental agreements: The case of global and transboundary pollution. In H. Folmer & T. Tietenberg (Eds.), *International yearbook of environmental and resource economics 2003/04* (pp. 82–158). Cheltenham: Edward Elgar.

Gillies, D. B. (1953). *Some theorems on N-person Games.* Ph.D. dissertation, Department of Mathematics, Princeton University.

Gordon, H. S. (1954). The economic theory of a common property resource: The fishery. *Journal of Political Economy, 62*(2), 124–142.

Kaitala, V., & Lindroos, M. (1998). Sharing the benefits of cooperation in high seas fisheries: a characteristic function game approach. *Natural Resource Modeling, 11*(4), 275–299.

Kronbak, L., & Lindroos, M. (2007). Sharing rules and stability in coalition games with externalities. *Marine Resource Economics, 22*(2), 137–154.

Lindroos, M., & Kaitala, V. (2000). Nash equilibria in a coalition game of the Norwegian spring-spawning herring fishery. *Marine Resource Economics, 15,* 321–340.

Lindroos, M., Kronbak, L., & Kaitala, V. (2007). Coalitions in fisheries games. In T. Bjørndal, D. Gordon, R. Arnason, & U. Sumaila (Eds.), *Advances in fisheries economics—Festschrift in honour of Professor Gordon R. Munro* (pp. 184–195). Colorado: Blackwell.

Mesterton-Gibbons, M. (2000). *An introduction to game-theoretic modelling.* Providence: American Mathematical Society.

Munro, G. (2000). The UN Fish Stocks Agreement of 1995: History and problems of implementation. *Marine Resource Economics, 15,* 265–280.

Schaefer, M. B. (1954). Some aspects of the dynamics of populations important to the management of the commercial marine fisheries. *Bulletin Inter-American Tropical Tuna Commission, 1,* 27–56.

Shapley, L. (1953). A value for n-person games. In H. W. Kuhn & A. W. Tucker (Eds.), *Contributions to the theory of games* (Vol. 2). Princeton, NJ: Princeton University Press.

United Nations. (1995). *United Nations Conference on Straddling Fish Stocks and Highly Migratory Fish Stocks. Agreement for the Implementation of the Provisions of the United Nations Convention on the Law of the Sea of 10 December 1982 Relating to the Conservation and Management of Straddling Fish Stocks and Highly Migratory Fish Stocks.* UN Doc. A/Conf./164/37.

Chapter 6
Non-cooperative Coalition Formation Games in Fisheries

Abstract Coalition formation games, in particular partition function games (p-games), have become a standard tool to understand the formation and stability of international fisheries agreements. This chapter presents a stylized p-game to model internationally shared fisheries. In a p-game, the formation of coalitions is endogenous, as a result of a non-cooperative process. The main outcomes of this model are discussed, namely, the free-rider incentives and the success of cooperation. This is followed by two case studies, which show different levels of cooperation among the harvesting countries: the East Atlantic and Mediterranean bluefin tuna, and the Norwegian spring spawning herring. Finally, policy implications for the management of internationally shared fish stocks are discussed. As shown in Chaps. 4 and 5, cooperative games are about the sharing of the aggregate worth of a given coalition among its members. These games leave unanswered an important question: which coalition will form? Coalition formation games provide the answer to this question by determining the equilibrium coalition endogenously. In this chapter, we concentrate on the most widely used type of coalition formation games: partition function games (p-games).

6.1 p-Game: The Concept

A partition function game $\Gamma(N, \Pi)$ is fully defined by a set of players, $N = \{1, 2, \ldots, n\}$, and a partition function, Π. In order to define partition function, we need to introduce the following concept.

Definition 6.1 A **coalition structure** $C = \{S_1, S_2, \ldots, S_z\}$ is a partition of the set of players, that is: $\bigcup_{k=1}^{z} S_k = N$ and $S_i \cap S_j = \emptyset, \forall i, j \in \{1, \ldots, z\} \wedge i \neq j$.

L. Grønbæk et al., *Game Theory and Fisheries Management*,
https://doi.org/10.1007/978-3-030-40112-2_6

107

A coalition structure is thus a partition of the set of players, that is, a set of coalitions that altogether integrates all the players, with each player belonging to only one coalition. Thus, it is the way in which all the players distribute themselves into coalitions. Following Bloch (2003), partition function can now be defined.

Definition 6.2 A **partition function** Π is a mapping which associates to each coalition structure $C = \{S_1, S_2, \ldots, S_z\}$, a vector in $\Re^{|C|}$, representing the worth of all coalitions in C.

A partition function yields the worth of all coalitions, for each partition of the set of players. In a partition function, the worth of a coalition S_k depends on the overall coalition structure: $\Pi_{S_k}(C)$. Thus, it is a generalization of the characteristic function, in which the worth of any coalition does not depend on how the remaining players form coalitions.

The worth of a coalition is generally affected by the merger of other coalitions. Hence, mergers generate positive or negative externalities for non-members. The partition function captures these externalities across coalitions, which are assumed to be absent in the characteristic function. Hence, as argued by Pintassilgo et al. (2015), the advantage of the partition function approach, compared to characteristic function games, is that it captures externalities across players compactly and allows one to analyse also the formation and stability of sub-coalitions. Fisheries games generally exhibit positive externality games as when players join into coalitions they usually find it optimal to decrease their aggregate fishing effort, which increase the payoff of the remaining players.

6.2 A Standard p-Game

This section presents a standard p-game applied to fisheries, introduced by Pintassilgo and Lindroos (2008). This game approaches the management of an internationally shared fish stock assuming *ex ante* symmetric players, that is, the fishing fleets have the same harvesting costs and sell the fish at the same price. This game assumes the classical Gordon–Schaefer bioeconomic model, introduced in Chap. 3 (Gordon 1954; Schaefer 1954), and a coalition formation model, composed of two stages: the 1st stage in which players decide whether to join a coalition or not, and the 2nd stage in which they decide on their fishing efforts.

6.2.1 The Bioeconomic Model

Recall the Gordon–Schaefer model presented in Chaps. 3 and 5. For a given set of players, $N = \{1, \ldots, n\}$, the relation between the fish stock, X, the harvest of an individual player i, H_i, and the fishing effort exerted by player i, E_i, is captured by the following three equations:

$$F(X) = rX\left(1 - \frac{X}{k}\right) \tag{6.1}$$

$$H_i = q E_i X \tag{6.2}$$

$$F(X) = \sum_{i=1}^{n} H_i, \tag{6.3}$$

where F denotes the natural growth function of the stock, r its intrinsic growth rate, k the carrying capacity of the ecosystem (the equilibrium level of X in the absence of harvesting), and q the catchability coefficient.

According to this static model, harvests take place on a sustainable yield basis, as the sum of harvests over all players equals the stock growth (Eq. 6.3). Upon substitution of Eqs. (6.1) and (6.2) into Eq. (6.3), the stock level can be expressed as function of the total fishing effort:

$$X = \frac{k}{r}\left(r - q \sum_{i=1}^{n} E_i\right). \tag{6.4}$$

This indicates a negative relation between the stock level and players' total fishing effort, $\sum_{i=1}^{n} E_i$.

The economic rent or payoff of fishing state i, Π_i, is defined as

$$\Pi_i = p H_i - c E_i, \tag{6.5}$$

where p is the price for fish and c the individual cost per unit of effort.

6.2.2 The Coalition Formation Model

Consider that players have to take two decisions: whether or not to join an international fisheries agreement (IFA) and the level of fishing effort. This is modelled using a two-stage framework.

The First Stage

In the first stage, each player decides on its participation: join the IFA (coalition) or be a non-member and act as a singleton. It is assumed that only one coalition (IFA) forms and that any player (country) is allowed to join it. The assumption of only one coalition comes from the UN Fish Stocks Agreement (United Nations 1995), according to which internationally shared fish stocks should be managed through Regional Fisheries Management Organizations (RFMOs). The assumption of open membership is based on Article 8 of the UN Fish Stocks Agreement, which states that participation in an RFMO should be open to all countries with "a real interest in the fisheries concerned" (see Sect. 1.2.1 for a discussion on these legal aspects).

Thus, first-stage decisions lead to a coalition structure, a partition of the set of players. This coalition structure is represented by $C = \{S, j_1, \ldots, j_{n-m}\}$ where $S = (i_1, \ldots, i_m)$ is a coalition (IFA) with m members, $m \in \{1, \ldots, n\}$, and $j_1, \ldots, j_{n-m}$ represent the $n - m$ singletons (non-members). In this context in which there is only one coalition, a coalition structure is fully characterized by that coalition, S.

Second Stage

In the second stage, players choose their fishing effort strategies. Assume that a coalition, S, has formed in the first stage. The steady-state payoff of this coalition is given by the sum of the payoffs of its members:

$$\Pi_S(S) = \sum_{i \in S} \Pi_i(S) = pqE_S \frac{k}{r}\left(r - qE_S - q\sum_{j \notin S} E_j\right) - cE_S, \qquad (6.6)$$

where $E_S = \sum_{i \in S} E_i$ stands for the aggregate effort of the coalition members, and E_j for the fishing effort of a non-member of coalition S.

The payoff of each non-member, j, is given by

$$\Pi_j(S) = pqE_j \frac{k}{r}\left(r - qE_S - qE_j - \sum_{\ell \notin \{S \cup j\}} E_\ell\right) - cE_j. \qquad (6.7)$$

The problem of coalition S is to maximize the sum of the payoffs of its members, $\text{Max}_{E_S} \Pi_S(S)$, and each non-member, j, maximizes its own payoff, $\text{Max}_{E_j} \Pi_j(S)$. The solution of the second-stage game is obtained by solving both problems simultaneously, which yields the Nash equilibrium fishing efforts:

$$E_S^*(S) = E_j^*(S) = \frac{r(1 - b)}{(n - m + 2)q}, \qquad (6.8)$$

where $b = \frac{c}{pqk}$, which always lies in the range $[0; 1]$,[1] is the "inverse efficiency parameter" presented in Chaps. 3 and 5.

---

[1] This can be shown by noting that the equilibrium stock levels in the open access regime is given by $X^{OA} = c/pq$, which is obtained by substituting Eq. (6.2) into Eq. (6.5) and setting profits to

Thus, the aggregate fishing effort is

$$AE(S) = E_S^*(S) + (n - m)E_j^*(S) = \frac{(n - m + 1)r}{(n - m + 2)q}(1 - b). \qquad (6.9)$$

The equilibrium stock level is obtained by inserting the aggregate fishing effort in Eq. (6.9) into Eq. (6.4):

$$X^* = k\left(1 - \frac{(n - m + 1)}{(n - m + 2)}(1 - b)\right). \qquad (6.10)$$

Inserting the equilibrium fishing efforts into the countries' payoffs, for each coalition structure, yields the equilibrium payoffs. Since the countries are symmetric, we assume an equal sharing of the coalitional worth. The equilibrium payoffs of coalition members ($i \in S$) and singleton ($j \notin S$) are given by

$$\Pi_{i \in S}^*(S) = \frac{rpk}{m(n - m + 2)^2}(1 - b)^2 \qquad (6.11)$$

$$\Pi_{j \notin S}^*(S) = \frac{rpk}{(n - m + 2)^2}(1 - b)^2. \qquad (6.12)$$

Solution of the Two-Stage Game

In each stage, strategies (participation and fishing effort) form a Nash equilibrium. The game is solved backward for the subgame perfect equilibrium. First, the Nash equilibrium fishing efforts of the second stage are computed, for each possible coalition structure. These fishing efforts are then used to obtain the payoffs for all possible coalitions (the partition function). Finally, the equilibrium participation strategies of the first stage are determined. These strategies also form a Nash equilibrium. This is characterized by the absence of incentives to change the participation decision, both for RFMO members (internal stability) and non-members (external stability).

Definition 6.3 Internal stability holds for a coalition S if $\Pi_i(S) \geq \Pi_i(S \setminus \{i\}) \ \forall i \in S$. **External stability** holds for a coalition S if $\Pi_j(S) \geq \Pi_j(S \cup \{j\}) \ \forall j \notin S$.

Thus, no member of S should have an incentive to leave the coalition (internal stability) and no non-member should have an incentive to join it (external stability). In the first stage, a coalition S is a Nash equilibrium if it is internally and externally stable.

---

zero. Thus, the inverse efficiency parameter can be written as $b = X^{OA}/k$. This ratio always lie in the range [0; 1] as the equilibrium stock level in open access lies between a minimum of zero (depletion of the stock) and a maximum of the carrying capacity of the ecosystem (no harvest), k.

6.2.3 Properties of the Game

Using the equilibrium payoffs, Eqs. (6.11) and (6.12), the following properties related to the second stage can be established.

Proposition 6.1 Impact of Enlarging an RFMO. *Consider a n-player game. Suppose a non-member $j \notin S$ joins the RFMO such that $S' = S \cup \{j\}$.*

(i) Positive Externalities: The payoff of country ℓ, who is neither a member of coalition S nor of S', is strictly higher under S' than under S: $\Pi_{\ell \notin S'}(S') > \Pi_{\ell \notin S}(S)$;

(ii) Global Efficiency from Cooperation: The aggregate payoff (the sum of the payoffs over all players) is strictly higher under S' than S:
$$\sum_{i \in S'} \Pi_i(S') + \sum_{\ell \notin S'} \Pi_\ell(S') > \sum_{i \in S} \Pi_i(S) + \sum_{\ell \notin S} \Pi_\ell(S).$$

The proof of this and the following proposition is shown in the appendix. Proposition 6.1, based upon Proposition 1 in Pintassilgo and Lindroos (2008), establishes that the game presents positive externalities, as the payoff of a singleton increases when another player joins the RFMO. Positive externalities are a key force that inhibits the formation of large coalitions. Global efficiency from cooperation implies that the aggregate payoff increases with the number of countries joining the RFMO. Thus, more cooperation produces larger aggregate payoffs.

6.2.4 Solution of the Game

Inserting the equilibrium payoffs, Eqs. (6.11) and (6.12), into the internal and external stability conditions yields the equilibrium of the first-stage game, which is shown in Proposition 6.2.

Proposition 6.2 Equilibrium coalition structure. If $n = 2$, then the equilibrium coalition structure is the Grand Coalition, $C = \{N\}$.

If $n \geq 3$ the equilibrium coalition structure is full non-cooperation, in which all players act as singletons, $C = \{1, 2, \ldots, n\}$.

Proposition 6.2, based upon Propositions 2 and 3 in Pintassilgo and Lindroos (2008), indicates pessimistic prospects regarding the cooperative management of internationally shared stocks under RFMOs. If the number of fishing countries is 3 or more, full non-cooperation is the equilibrium outcome.

6.2.5 A Numerical Example

This section provides a numerical example to illustrate the outcomes of a game with two and three symmetric players. The following parameter values are considered: $r = 1; k = 100; q = 0.004; p = 3;$ and $c = 0.3$.

In the two-player game, the set of players is given by $N = \{1, 2\}$. There are two possible coalition structures, the one in which players 1 and 2 form a coalition, which is represented as $C_1 = \{(1, 2)\}$, and the one in which they behave as singletons, represented as $C_2 = \{1, 2\}$.

When the Grand Coalition forms, the fishing effort level of the coalition (Eq. 6.8) is $E_S^* = 93.75$. The steady-state stock level (Eq. 6.10) is $X^* = 62.5$ and, assuming an equal sharing of the coalition payoff, each player receives the payoff $\Pi_{i \in S}^* = 21.09$ (Table 6.1).

If players behave non-cooperatively their fishing effort levels are $E_1^* = E_2^* = 62.5$, leading to an aggregate effort level, $AE = 125$, higher than under cooperation and hence a lower stock level, $X^* = 50$. The payoff of each player, $\Pi_j^* = 18.75$, is smaller than the one obtained under cooperation. Thus, cooperation is the Nash equilibrium of the game, as players do not have incentive to deviate.

Consider now the case of three symmetric players. The set of players is now given by $N = \{1, 2, 3\}$. The Grand Coalition is represented by $C_1 = \{(1, 2, 3)\}$, and there are three coalition structures with 2-player coalitions, $C_2 = \{(1, 2), 3\}$, $C_3 = \{(1, 3), 2\}$, and $C_4 = \{(2, 3), 1\}$. Full non-cooperation is represented by $C_5 = \{1, 2, 3\}$. The payoffs of coalition members and non-members, for all coalition structures, are shown in Table 6.2.

The payoffs in Table 6.2 show that the Grand Coalition is not internally stable, as any player can increase its payoff, from 14.06 to 18.75, by leaving the cooperative agreement and adopting a free-rider behaviour. Two-player coalitions are also not stable as their members can increase their payoffs, from 9.38 to 10.55, by leaving the

Table 6.1 Payoffs of a two-player game

Coalition structure[a]	Payoff of coalition member	Payoff of non-member
$\{(1, 2)\}$	21.09	–
$\{1, 2\}$	–	18.75

[a]Players inside round brackets represent a coalition

Table 6.2 Payoffs of a three-player game

Coalition structure	Coalition member	Non-member
$\{(1, 2, 3)\}$	14.06	–
$\{(1, 2), 3\}$ or $\{(1, 3), 2\}$ or $\{(2, 3), 1\}$	9.38	18.75
$\{1, 2, 3\}$	–	10.55

coalition. The only stable coalition structure is the one formed by singletons, which is externally stable, as players do not have incentive to form a two-player coalition, and also internally stable, as no further deviations are possible.

6.3 Extension: The Role of Asymmetry

In this section, the assumption of symmetric players is relaxed, as in Pintassilgo et al. (2010), by allowing players to have different costs per unit of fishing effort. Thus, the payoff of player i, Π_i, is now given by

$$\Pi_i = pH_i - c_i E_i, \tag{6.13}$$

where c_i the individual cost per unit of effort of player i.

6.3.1 The Coalition Formation Model

Under cost asymmetry, each coalition S chooses to allocate its fishing effort to the member will the lowest cost per unit of effort, which we denote by $c_S^{\min}$. Thus, the payoff of coalition S is given by

$$\Pi_S(S) = \sum_{i \in S} \Pi_i(S) = pqE_S \frac{k}{r} \left(r - qE_S - q \sum_{j \notin S} E_j \right) - c_S^{\min} E_S. \tag{6.14}$$

The payoff of each non-member, j, is given by

$$\Pi_j(S) = pqE_j \frac{k}{r} \left(r - qE_S - qE_j - \sum_{\ell \notin \{S \cup j\}} E_\ell \right) - c_j E_j. \tag{6.15}$$

Both coalition S and each non-member, j, choose the fishing effort strategies that maximize their payoffs: $\text{Max}_{E_S} \Pi_S(S)$ and $\text{Max}_{E_j} \Pi_j(S)$, respectively. Solving these problems simultaneously, and assuming interior solutions, we obtain the following equilibrium fishing effort levels:

$$E_S^*(S) = \frac{(n-m+1)r}{(n-m+2)q}\left(1 - b_S^{\min}\right) - \frac{r}{(n-m+2)q}\sum_{j \notin S}(1 - b_j) \tag{6.16}$$

$$E_j^*(S) = \frac{(n-m+1)r}{(n-m+2)q}(1-b_j) - \frac{r}{(n-m+2)q}\left[(1-b_S^{\min}) + \sum_{k \neq j \notin S}(1-b_k)\right],$$
(6.17)

where $b_S^{\min} = \frac{c_S^{\min}}{pqk}$ and $b_j = \frac{c_j}{pqk}$.

By inserting the equilibrium effort levels into the payoffs in Eqs. (6.14) and (6.15), we obtain the equilibrium payoffs of coalition S and singleton j:

$$\Pi_S^*(S) = \frac{rpk}{(n-m+2)^2}\left(1 - (n-m+1)b_S^{\min} + \sum_{j \notin S} b_j\right)^2$$
(6.18)

$$\Pi_{j \notin S}^*(S) = \frac{rpk}{(n-m+2)^2}\left(1 - (n-m+1)b_j + b_S^{\min} + \sum_{k \neq j \notin S} b_k\right)^2.$$
(6.19)

To analyse the stability of a coalition in a context of asymmetric players, it is useful to introduce the concept of potential internal stability (PIS).

Definition 6.4 Potential internal stability hold for a coalition S if $\Pi_S(S) \geq \sum_{i \in S} \Pi_i(S \setminus \{i\})$, where $\Pi_S(S)$ represents the payoff of coalition S, and $\Pi_i(S \setminus \{i\})$ the payoff of player i after leaving coalition S.

A coalition S is potentially internally stable if its payoff is larger or equal than the sum of the payoffs of its members when leaving the coalition, that is, their free-rider payoffs. In other words, when the surplus of the coalition payoff over the sum of the members' free-rider payoffs is non-negative, the coalition is potentially internally stable. Thus, when PIS holds there is always a way to share the coalition worth such that no member wants to leave. Consequently, if a coalition S is PIS then there is always a sharing rule that makes it internally stable.

Under asymmetry, the rule for sharing the coalition worth is a central issue. The equal sharing assumed under symmetry is no longer the unique option. As shown in Chap. 5, several sharing rules have been proposed, such as the Nash bargaining solution, the Shapley value and the nucleolus. Herein we present another rule, the "almost ideal sharing scheme" (AISS) proposed by Eyckmans and Finus (2009). According to the AISS, the coalitional payoff is shared among its members such that each one receives his free-rider payoff, $\Pi_i(S \setminus \{i\})$, plus a share $\lambda_i(S)$ of the Coalition Surplus $\Delta(S)$:

$$\Pi_i(S) = \Pi_i(S \setminus \{i\}) + \lambda_i(S)\Delta(S),$$
(6.20)

where $\Delta(S) = \Pi(S) - \sum_{i \in S} \Pi_i(S \setminus \{i\})$, $\sum_{i \in S} \lambda_i(S) = 1$, $\lambda_i(S) > 0, \forall i \in S$.

From the definitions of PIS and AISS, it is evident that if a coalition S is PIS, then the AISS makes S internally stable, irrespective of the weights λ_i. Thus, the AISS mitigates the incentives for leaving a coalition. A second property of the AISS is that it allows establishing a direct link between internal and external stability: if coalition S is PIS, then all coalitions $S\backslash\{i\}$ for all $i \in S$ are not externally stable, irrespective of weights. Thus, under the AISS, external stability can be inferred directly from PIS. A third property of the AISS is that the internally and externally stable coalitions do not depend on the weights adopted.

Contrary to the case of symmetry, in which the equilibrium of a game with three or more players is full non-cooperation, under some types of cost asymmetry it is possible to stabilize large coalitions and even the Grand Coalition, as shown in the example below.

6.3.2 A Numerical Example

Consider the three-player game in the numerical example shown in Sect. 6.2.5. Let us depart from cost asymmetry ($c = 0.3$) by setting $c_1 = 0.3$, $c_2 = 0.3$, $c_3 = 0.1$. All the other parameter values remain unchanged: $r = 1$; $k = 100$; $q = 0.004$; $p = 3$.

The payoffs of the coalition and non-members, computed based on Eqs. (6.18) and (6.19), are shown in Table 6.3, for all possible coalition structures. The Coalition Surplus, as defined in Eq. (6.20), is also computed. Then, assuming that the AISS is adopted to share the coalition payoff among its members, the internal and external stability is analysed. A non-negative Coalition Surplus implies that the coalition is PIS and hence internally stable under the AISS. If the Coalition Surplus is negative, then the coalition is not PIS and consequently not internally stable.

In this game, the Grand Coalition is the unique stable coalition. It is externally stable by definition and also internally stable under the AISS as it presents a positive Coalition Surplus. Since the Grand Coalition is internally stable, all two-player coalitions are not externally stable as the free riders have incentive to join the coalitions

Table 6.3 Payoffs of a three-player game

Coalition structure[a]	Coalition	Non-member	Coalition Surplus	Internal stability	External stability
$\{(1, 2, 3)\}$	63.02	–	1.22	Yes	Yes
$\{(1, 2), 3\}$	11.34	39.12	−1.42	No	No
$\{(1, 3), 2\}$	39.12	11.34	3.44	Yes	No
$\{(2, 3), 1\}$	39.12	11.34	3.44	Yes	No
$\{1, 2, 3\}$	–	\{6.38, 6.38, 29.30\}	0	Yes	No

[a]Players inside round brackets represent a coalition

under the AISS. Finally, the singleton coalition structure is not externally stable as players 1 and 3, as well as 2 and 3, have incentive to join.

6.3.3 Simulation Results

Considering the full range of parameter values, it is possible to compute, through simulations, the likelihood of a given coalition S of size m being stable in a n-player game. For this purpose, let us insert the equilibrium payoffs (Eqs. 6.18 and 6.19) into the PIS condition (Definition 6.4). This yields

$$\left(\frac{n-m+3}{n-m+2}\right)\left(1-(n-m+1)b_S^{\min}+\sum_{j\notin S}b_j\right)^2 \geq$$
$$\sum_{i\in S}\left(1-(n-m+2)b_i+b_{S'}^{\min}+\sum_{k\neq i\notin S'}b_k\right)^2,$$
(6.21)

where $S' = S\setminus\{i\}$ and $b_{S'}^{\min}$ is the lowest b in coalition S'. The interesting aspect of this condition is that, for a given number of players n, and coalition size m, potential internal stability only depends on vector $b = (b_1, \ldots, b_n)$.

Assuming that parameters $b_1, \ldots, b_n$ are uniformly and independently distributed in the admissible range $[0, 1]$, Pintassilgo et al. (2010) used Monte Carlo simulations to compute the proportion of cases in which a coalition of size m is stable in a n-player game. This proportion, which can be understood as a probability, or likelihood, was computed based on the PIS condition (6.21), assuming that the coalition worth is shared among the members according to the AISS (6.20).

A given coalition S is stable if and only if it is internally stable and externally stable (Definition 6.3). As referred above, if the PIS condition holds for a coalition S, then, under the AISS, that coalition is internally stable. Moreover, external stability can also be inferred from the PIS condition.

To analyse the success of cooperation, the closing the gap index (CGI) is particularly useful. It measures how much the stable equilibria succeed in closing the gap between full cooperation and no cooperation. For a given stable coalition $\left(S_j^*\right)$, this index is defined as

$$\text{CGI}\left(S_j^*(b),n\right) = \frac{A\Pi\left(S_j^*\right)-A\Pi(\{1,2,\ldots,n\})}{A\Pi(N)-A\Pi(\{1,2,\ldots,n\})} \times 100,$$
(6.22)

where $A\Pi\left(S_j^*\right)$ represents the aggregate payoff of all players when the stable coalition S_j^* forms, which is obtained by summing the coalition worth to the payoffs of non-members. $A\Pi(N)$ represents the aggregate payoff under the Grand Coalition and $A\Pi(\{1,2,\ldots,n\})$ the aggregate payoff when all players act as singletons.

Table 6.4 Stability likelihood

		Number of players (n)								
		2	3	4	5	6	7	8	9	10
Number of coalition members (m)	n	1	0.451	0.051	0.001	0	0	0	0	0
	$n-1$	0	0.298	0.146	0.031	0.005	0.001	0	0	0
	$n-2$	–	0.084	0.251	0.100	0.025	0.004	0.001	0	0
	$n-3$	–	–	0.092	0.185	0.071	0.018	0.003	0	0
	$n-4$	–	–	–	0.068	0.138	0.050	0.013	0.002	0
	$n-5$	–	–	–	–	0.049	0.116	0.038	0.009	0.002
	$n-6$	–	–	–	–	–	0.036	0.098	0.032	0.007
	$n-7$	–	–	–	–	–	–	0.029	0.088	0.025
	$n-8$	–	–	–	–	–	–	–	0.024	0.079
	$n-9$	–	–	–	–	–	–	–	–	0.021
$\overline{\text{CGI}}(n)$		100	67.8	31.6	17.6	11.1	7.3	5.0	3.6	2.6

Source Adapted from Pintassilgo et al. (2010, 392)

Table 6.4 shows the stability likelihood of coalitions of size $m \in \{1, 2, \ldots, n\}$, for games from 2 to 10 players, based on 50,000 simulations. The average value of the CGI over the simulations $\overline{(\text{CGI})}$ is also presented.

The results presented in Table 6.4 show that the larger the number of fishing states involved in a fishery, the lower is the likelihood of a large coalition being stable. In particular, the Grand Coalition, $m = n$, is a likely outcome only in fisheries with a low number of fishing states. The closing the gap index (CGI) decreases with the number of players. This indicates that the larger the number of players, the lower the success of cooperation.

In order to assess the role of cost asymmetry in the success of cooperation, three ranges for parameter b_i, $i = 1, \ldots, n$, were considered, with the same mean value: $\{0.2\}$, $[0.1, 0.3]$ and $[0, 0.4]$. The first corresponds to a case of symmetry in which $b_1 = \ldots = b_n = 0.2$. In the second, $b_1, \ldots, b_n$ are uniformly and independently distributed in the range $[0.1, 0.3]$, which generates asymmetric values. Finally, in the third, parameters are distributed over a larger range, thereby increasing asymmetry. Table 6.5 shows the value of the average CGI for these three ranges.

Table 6.5 Coalition gain index: the asymmetry effect

Range of b_i^s	Number of players								
	2	3	4	5	6	7	8	9	10
0.2	100	0.0	0.0	0.0	0.0	0.0	0.0	0.0	0.0
[0.1; 0.3]	100.0	43.8	17.3	9.9	6.6	4.7	3.5	2.7	2.1
[0; 0.4]	100.0	76.7	42.7	23.3	13.5	8.4	5.5	3.9	2.8

Source Adapted from Pintassilgo et al. (2010, 394)

Table 6.6 Coalition gain index: the efficiency effect

Range of b_i^s	Number of players								
	2	3	4	5	6	7	8	9	10
[0; 0.2]	100	38.9	15.2	8.5	5.7	4.1	3.1	2.4	1.9
[0.2; 0.4]	100.0	50.2	19.8	11.4	7.6	5.4	4.0	3.0	2.3
[0.4; 0.6]	100	67.8	31.6	17.6	11.1	7.3	5.0	3.6	2.6

Source Adapted from Pintassilgo et al. (2010, 394)

Table 6.5 shows that, for a given number of players, CGI increases with the degree of asymmetry. That is, the larger the asymmetry the larger is the success of cooperation. This occurs, as asymmetry allows a cost-effective allocation of the fishing effort within the coalition, therefore increasing the gains from cooperation.

Another important aspect to analyse is the effect of level of efficiency, measured through parameter b_i, on the success of cooperation. To isolate the effect of efficiency, consider three ranges for b_i with different mean values but equal amplitude and hence asymmetry: [0; 0.2], [0.2; 0.4] and [0.4; 0.6]. The mean value of b_i rises from 0.1 in the first range to 0.3 in the second and 0.5 in the third. The results of the average CGI are shown in Table 6.6.

Table 6.6 shows that CGI increases with the average value of b_i. As this is an "inverse efficiency parameter", we can conclude that the success of cooperation decreases with the efficiency level. As $b_i = c_i/pqk$, a decrease in this parameter, that is, an increase in efficiency, can be due to a decrease in the cost per unit of effort, or an increase in price, catchability or carrying capacity of the ecosystem. The changes of these parameters in that direction make cooperation less successful.

The above partition game has been further extended to consider Stackelberg leadership (Long and Flaaten 2011), marine protected areas (Punt et al. 2013) and non-consumptive values (Pintassilgo et al. 2018). A partition function game has also been applied to model an internationally shared fishery, incorporating the spatial dimension of the stock (Finus et al. 2020). This application is discussed in the next chapter.

6.4 Case Study: The Eastern Atlantic and Mediterranean Bluefin Tuna Fishery

The bluefin tuna fishery in the East Atlantic illustrates well the lessons provided by p-games. The bluefin is the largest and most valued of the Atlantic tunas. Being managed under the aegis of a Regional Fisheries Management Organization (RFMO), the International Commission for the Conservation of Atlantic Tunas (ICCAT), it has been pointed out as a case of low cooperation among the harvesting countries.

The bluefin is a highly migratory fish stock widely distributed in the East Atlantic, from the Canary Islands to Norway, in the North Sea and in all the Mediterranean

Sea, where its spawning areas are located. A large number of countries is involved in this fishery, 24 according to ICCAT (2016), using a variety of fishing gears (e.g. bait boat, longline, purse seine, traps and sport gears).

During the period 1975–1990 the annual catches of the bluefin tuna in the East Atlantic and Mediterranean lied below 25,000 t (ICCAT 2000). The catches more than doubled, to values between 50,000 and 61,000 t, from mid-1990s to 2007 (ICCAT 2016). This period was characterized by major under-reporting of catches to the ICCAT and lack of compliance with the total allowable catches set by this organization. As a consequence, the stock declined sharply: by mid-2000s the spawning stock biomass was about half of its value in the early 1970s (ICCAT 2016).

The lack of cooperation in the bluefin tuna fishery is according to the lessons provided by p-games. This fishery is characterized by a large number of harvesting countries (players) and a high price (hence low $b_i = c_i/pqk$ in our model). As shown in the previous section, both elements lead to low success of cooperation in a p-game with asymmetric players. This was confirmed by Pintassilgo (2003) who applied a p-game to the bluefin tuna fishery in the East Atlantic. Using a fairly disaggregated bioeconomic model, including the age structure of the fish stock and multi-gears, the author concluded that the Grand Coalition was not stable as there was no sharing rule that could stabilize it.

Faced with the collapse of the bluefin tuna stock in the East Atlantic and Mediterranean by mid-2000s and the non-cooperative incentives of the harvesting countries, the ICCAT reacted by limiting free-rider strategies and enforcing strict regulations to reduce total harvest. A recovery plan was implemented, which included monitoring and enforcement controls to limit illegal, unreported and unregulated fishing. The total allowable catch was significantly reduced, from 32,000 t in 2008 to 13,500 t in 2010 (ICCAT 2011). As a result, since 2008 a substantial decrease in the bluefin tuna catch occurred in the Eastern Atlantic and Mediterranean Sea and the stock started recovering.

6.5 Case Study: The Norwegian Spring Spawning (Atlanto-Scandian) Herring

The Norwegian spring spawning, or Atlanto-Scandian, herring resource has historically been one of the most important and valuable fishery resources in the North Atlantic. It has been of particular importance to Norway for centuries (Sandberg 2010). At its peak production in the mid-1960s, the fishery yielded harvests (albeit unsustainable) of 2 million tonnes (MT) per annum. When in a healthy state, the resource migrates from Norway to Iceland, and pre-Brexit was subject to exploitation by Norway, Iceland, Russia, the Faroe Islands and the EU. It has been, since the advent of the EEZ regime, a straddling stock as it passes through a high seas enclave, the Banana Hole (Bjørndal 2009).

The history of the fishery during the first three decades after the Second World War was similar in many respects to that of North Sea herring, discussed in Chap. 4. ICES scientists estimate that the safe minimum spawning stock biomass (SSB) is 2.5 MT (Bjørndal 2009). At the end of the Second World War, the SSB was far above that level, as fishing for Norwegian spring spawning herring during the War had not been without its risks.

North Sea herring was described as a resource that is vulnerable to overfishing, because harvesting costs are relatively insensitive to the size of the biomass. This is equally true of the Norwegian spring spawning herring. Nonetheless, the resource seemed secure until the 1960s. To a considerable extent, this was due to the fact that the resource extends far out into the Atlantic, which meant that given the fishing technology of the day a significant part of the resource was shielded from extensive exploitation by economics.

Rapid technological advances in the fishing of herring in the late 1950 and 1960s stripped away the economic protection. This stripping away of the economic protection came to be combined with adverse environmental conditions, with the consequence that by the late 1960s, serious concern was being expressed about the state of the resource (Sandberg 2010). Like North Sea herring, the Norwegian spring spawning herring resource was ostensibly under the management control of the Northeast Atlantic Fisheries Commission (NEAFC). Also like North Sea herring, the ability of NEAFC to take corrective action proved to be weak and ineffective. By the early 1970s, the resource had crashed.

To repeat, the minimum safe Norwegian spring spawning herring SSB was and is, according to ICES, 2.5 MT. It is estimated that by the early 1970s, the SSB had plummeted to 2 thousand tonnes, less than one-tenth of 1% of the minimum level (Bjørndal 2008). The resource had come within a hair's breadth of extinction. The remnants of the resource did not extend far from the shores of Norway. With the coming of the EEZ regime in Europe, the remnants of the Norwegian spring spawning herring resource were found to be contained solely within the Norwegian EEZ, and thus wholly under Norwegian control. The Norwegians implemented a near harvest moratorium that was to remain in place until the mid-1990s.[2]

The near harvest moratorium, along with a return to favourable oceanic conditions, resulted in an eventual recovery of the resource (Miller et al. 2013). By the mid-1990s, there was evidence that the resource was resuming its migratory pattern. The near harvest moratorium was lifted. The resource was once more open to exploitation by Iceland, Russia, the Faroe Islands and the EU, as well as Norway.

The five needed no lectures on the Prisoner's Dilemma. After some initial stumbling, the five came together to form a cooperative resource management arrangement—a stable fisheries cooperative game. They were fortunate in that UN Fish

---

[2]When the resource is healthy, the herring migrate in search of food. No such search was required by the survivors. The Norwegians allowed a small local fishery, but with a strict ban on small herring, which in the past had accounted for a third of the Norwegian catch (Sandberg 2010).

Stocks Agreement had made its appearance in 1995, thereby providing a legal framework. The five parties did and do constitute a de facto RFMO (see Munro 2011, for details).

The cooperative resource management regime followed scientific advice from ICES in setting the global TAC. The harvest shares were of the fixed through time form, described in Chap. 4. They, like the harvest shares for North Sea herring, were and are based upon the so-called zonal attachment, that is to say, based upon the amount of time per year the resource was estimated to spend in each of the respective five zones.

Side payment-like arrangements were implemented through bilateral agreements, between the lead player, Norway and the other four. The other four were granted the right to take parts of their quotas in the Norwegian EEZ and to land parts of their catches in Norway for processing. By allowing the other four to take parts of their quotas in the Norwegian EEZ, it meant that the other four would be catching herring when the fish were at their most valuable state. Allowing the other four to land their catches in Norway had the consequence of reducing transportation costs. The side payment-like arrangements were thus unquestionably Pareto improving, increasing the cooperative surplus (Miller et al. 2013).

For several years, the cooperative management regime worked very well. In 2000, one of the authors held up the Norwegian spring spawning herring management regime as an exemplary model of a well working RFMO (Munro 2000). The cooperative fishery game exhibited all of the signs of internal stability. Enhancing the internal stability, undoubtedly, was the fact that the number of players was small—five, a very small number for an RFMO. Furthermore, there was no evidence of unregulated fishing, nor did the new member problem arise. The fact that two out of the five players, the EU and Russia, were powerful politically most certainly helped in the regard.

In spite of the stability of this cooperative fishery game, two years after Munro put forth Norwegian spring spawning herring as an exemplary RFMO, the cooperative management regime was to provide an example of the time consistency problem. To re-phrase the matter, the cooperative management regime was to provide an example of an RFMO displaying a lack of resilience in the face of unexpected shocks.

The zonal attachment rule for harvest shares had been based upon the pre-crash migratory pattern of the resource, it being assumed that the restored resource would resume its former migratory pattern. By 2002, the resource migratory pattern estimates had been revised. The revised pattern turned out to be significantly different from that of the pre-crash era.

The revised estimates demonstrated, claimed Norway (the lead player), that its true zonal attachment was substantially greater than previously thought. Consequently, Norway argued, it should receive a substantially greater share of the global TAC. The other players responded by expressing admiration for the Norwegians' sense of humour. The five were at an impasse. By 2003, the cooperative management regime was in a state of paralysis; the hitherto internally stable cooperative fishery game degenerated into a competitive one (Miller et al. 2013).

The impasse continued for several dangerous years, with the Prisoner's Dilemma manifesting itself in two ways. First, commencing in 2003, each of the five players announced its own unilateral quota. By 2006, the sum of the unilateral quotas were exceeding by a significant margin the global quota implied by ICES advice (Bjørndal 2009). Second, in 2003, Norway announced that, with the exception of Russia, its bilateral agreements with its partners, pertaining to the harvesting and landing of herring catches, were to be suspended until further notice. This could be seen as a willful reduction of the Norwegian spring spawning herring global economic pie (Miller et al. 2013).

Eventually, a new cooperative management agreement was negotiated, with the negotiations coming to a conclusion in late 2006, and the new agreement being implemented at the beginning of 2007. Norway did receive an increased allocation, but the increase can only be described as modest. It can be conjectured that cooperation was restored as a consequence of the five gazing into the abyss that yawned before them. If the resource were to crash again, the crash might this time prove to be irreversible (Miller et al. 2013).

The restored fishery cooperative game has proven to be internally stable up to time of writing. By the end of the first decade of the new millennium, the SSB was at levels not seen since the 1950s (Sandberg 2010). The SSB is subject to natural fluctuations. The most recent estimates of the SSB shows it to be lower than what it was at the end of the previous decade, but still comfortably above the ICES declared safe minimum level (ICES 2018).

The Norwegian spring spawning herring cooperative resource management regime is not perfect. Both Sandberg (2010) and Bjørndal (2008) have argued that the resource rent arising from the fishery could be increased through improved management policies. The two willingly concede, however, that the resource rent arising is substantial and that it is at a level, which would have seemed beyond the realm of possibility in the early 1980s. Thus, as opposed to the bluefin tuna case study presented in the previous section, the Norwegian spring spawning herring can be considered as a successful case of cooperation among the harvesting countries.

6.6 Policy Implications

This chapter shows how coalition formation games can be used to analyse the formation and stability of international fisheries agreements (IFAs). The key result obtained is that large stable agreements are unlikely, due to the presence of free-rider incentives. That is, when many countries join an IFA, member countries have incentive to leave and behave non-cooperatively as singletons. Hence, large IFAs tend to be unstable. Addressing this problem requires measures to mitigate free-rider incentives by strengthening the role of Regional Fisheries Management Organizations (RFMOs) in regulating the fishing activities of both members and non-members. In particular, to deal with illegal, unreported and unregulated (IUU) fishing.

The likelihood of achieving a stable IFA incorporating all the fishing states, or a large majority of them, decreases significantly with the number of countries involved in the fishery. The reason is that the larger the number of countries, the less likely a coalition will be able to give all its members a payoff that covers the payoff they would earn by leaving the IFA to act non-cooperatively. Thus, those RFMOs with a large number of members should be especially active in preventing free riding. This result also implies that new entrants, that is, new countries joining a fishery, by increasing the number of fishing countries, make cooperation less likely. Hence, the RFMOs should establish mechanisms to regulate the fishing activities of new entrants, both in case they join the RFMO and in case they choose to be non-members.

The results also show that the success of cooperation decreases with the level of efficiency of the fishing fleets. This implies that the expected increase in the level of efficiency of fishing fleets in the future is likely to put extra pressure on the stability of RFMOs. Hence, these organizations should constantly find stronger mechanisms to deter free riding, as the incentives are likely to increase over time.

It is also found that asymmetry among fishing states can foster cooperation if appropriate transfer schemes are adopted. This shows that transfers between fishing states, or side payments, have the potential to strengthen the stability of RFMOs.

Appendix

Proof of Proposition 6.1

(i)   From Eq. (6.12), the payoff of player ℓ when coalition S forms is

$$\Pi^*_{\ell \notin S}(S) = \frac{rpk}{(n-m+2)^2}(1-b)^2. \qquad (6.23)$$

In the same way, the payoff of player ℓ when coalition $S' = S \cup \{j\}$ forms is

$$\Pi^*_{\ell \notin S'}(S') = \frac{rpk}{(n-(m+1)+2)^2}(1-b)^2. \qquad (6.24)$$

From Eqs. (6.23) and (6.24) it comes directly that $\Pi_{\ell \notin S'}(S') > \Pi_{\ell \notin S}(S)$, which proves that the game exhibits positive externalities.

(ii) Using Eqs. (6.11) and (6.12), we can obtain the aggregate payoff when coalition S forms:

$$\sum_{i \in S} \Pi_i(S) + \sum_{\ell \notin S} \Pi_\ell(S) = m\Pi_i(S) + (n-m)\Pi_\ell(S)$$

$$= m\frac{rpk}{m(n-m+2)^2}(1-b)^2 + (n-m)\frac{rpk}{(n-m+2)^2}(1-b)^2$$

$$= \frac{(n-m+1)rpk}{(n-m+2)^2}(1-b)^2. \qquad (6.25)$$

Following the same procedure, the aggregate payoff when coalition S' forms is

$$\sum_{i \in S'} \Pi_i(S') + \sum_{\ell \notin S'} \Pi_\ell(S') = (m+1)\Pi_i(S) + (n - (m+1))\Pi_\ell(S)$$

$$= (m+1)\frac{rpk}{(m+1)(n-(m+1)+2)^2}(1-b)^2 + (n-(m+1))\frac{rpk}{(n-(m+1)+2)^2}(1-b)^2$$

$$= \frac{(n-m)rpk}{(n-m+1)^2}(1-b)^2. \tag{6.26}$$

To conclude the proof, we just need to show that the aggregate payoff in Eq. (6.26) is larger than the one in Eq. (6.25). This condition is given by

$$\frac{(n-m)rpk}{(n-m+1)^2}(1-b)^2 > \frac{(n-m+1)rpk}{(n-m+2)^2}(1-b)^2. \tag{6.27}$$

The inequality (6.27) can be simplified to $\frac{(n-m)}{(n-m+1)^2} > \frac{(n-m+1)}{(n-m+2)^2}$ and further to $(n-m)^2 + (n-m) - 1 > 0$, which always holds since $m < m+1 \leq n$. Thus, global efficiency from cooperation holds. ∎

Proof of Proposition 6.2

According to Definition 6.3, the internal stability condition is

$$\Pi_i(S) \geq \Pi_i(S\setminus\{i\}) \quad \forall i \in S. \tag{6.28}$$

Using the payoffs in Eqs. (6.11) and (6.12), the internal stability condition can be written as

$$\frac{rpk}{m(n-m+2)^2}(1-b)^2 \geq \frac{rpk}{(n-(m-1)+2)^2}(1-b)^2, \quad m \geq 2. \tag{6.29}$$

This condition only applies for $m \geq 2$, as for $m = 1$ no further deviations are possible. Condition (6.29) can be simplified to

$$(1-m)(n-m)^2 + (6-4m)(n-m) + 9 - 4m \geq 0, \quad m \geq 2. \tag{6.30}$$

Following the same procedure, the external stability is given by

$$\frac{rpk}{(n-m+2)^2}(1-b)^2 \geq \frac{rpk}{(m+1)(n-(m+1)+2)^2}(1-b)^2, \quad m < n. \tag{6.31}$$

This condition only applies for $m < n$. When $m = n$, the Grand Coalition forms and thus there is no other player that could join it. In other words, the Grand Coalition is externally stable by definition.

Condition (6.31) can be simplified to

$$m(n - m)^2 + (2m - 2)(n - m) + m - 3 \geq 0, \quad m < n. \tag{6.32}$$

If $n = 2$, then the coalition structure formed by the two singletons $(m = 1)$ is not externally stable, condition (6.32) does not hold, and hence cannot be an equilibrium. The internal stability condition (6.30) holds for the Grand Coalition $(m = 2)$. As the Grand Coalition is externally stable by definition it is the equilibrium coalition structure.

If $n \geq 3$, then from condition (6.30) it is straightforward to conclude that for $m \geq 3$ internal stability does not holds (all the terms in the left-hand side of the inequality are non-positive). For $m = 2$, condition (6.30) also fails. In fact, it becomes $-n^2 + 2n + 1 \geq 0$, which never holds for $n \geq 3$. The coalition structure in which all players act as singletons $(m = 1)$ is internally stable by definition. This coalition structure is also externally stable, as for $m = 1$ condition (6.32) becomes $n^2 - 2n - 1 \geq 0$, which always holds for $n \geq 3$. Thus, the equilibrium coalition structure is the one formed by singletons. ∎

References

Bjørndal, T. (2008). *Rent in the fishery for Norwegian spring spawning herring*. Washington, DC: World Bank PROFISH Program.

Bjørndal, T. (2009). Overview, roles and performance of the Northeast Atlantic Fisheries Commission (NEAFC). *Marine Policy, 33,* 685–697.

Bloch, F. (2003). Non-cooperative models of coalition formation in games with spillovers. In C. Carraro (Ed.), *The endogenous formation of economic coalitions* (pp. 35–79). Cheltenham: Edward Elgar.

Eyckmans, J., & Finus, M. (2009). *An almost ideal sharing scheme for coalition games with externalities*. Stirling Economics Discussion Paper (2009–10).

Finus, M., Schneider, R., & Pintassilgo, P. (2020). Social and technical excludability of impure public agreements: The case of international fisheries. *Resource and Energy Economics,* 59.

Gordon, H. S. (1954). The economic theory of a common property resource: The fishery. *Journal of Political Economy, 62*(2), 124–142.

ICCAT. (2000). *Report for biennial period, 2000–01*, Part I (Vol. 2). Madrid: International Commission for the Conservation of Atlantic Tunas, SCRS.

ICCAT. (2011). *Report for biennial period, 2010–11*, Part I (Vol. 2). Madrid: International Commission for the Conservation of Atlantic Tunas, SCRS.

ICCAT. (2016). *Report for biennial period, 2016–17*, Part I (Vol. 2). Madrid: International Commission for the Conservation of Atlantic Tunas, SCRS.

ICES: International Council for the Exploration of the Seas. (2018). *Report of the working group on widely distributed stocks* (p. 23). Copenhagen: ICES CM2018/ACOM.

Long, L. K., & Flaaten, O. (2011). A Stackelberg analysis of the potential for cooperation in straddling stock fisheries. *Marine Resource Economics, 26*(2), 119–139.

Miller, K., Munro, G., Sumaila, U. R., & Cheung, W. (2013). Governing marine fisheries in a changing climate: A game-theoretic perspective. *Canadian Journal of Agricultural Economics, 61,* 309–334.

Munro, G. (2000). The United Nations Fish Stocks Agreement of 1995: History and problems of implementation. *Marine Resource Economics, 19,* 265–280.

Munro, G. (2011). *On the management of shared living marine resources.* Paper presented to the 2011 Conference of Danish Environmental Economic Council, Skodsberg, Denmark, 5–6 September 2011. Retrieved from http://www.dors.dk/sw8608.asp.

Pintassilgo, P. (2003). A coalition approach to the management of high seas fisheries in the presence of externalities. *Natural Resource Modeling, 16*(2), 175–197.

Pintassilgo, P., & Lindroos, M. (2008). Coalition formation in straddling stock fisheries: A partition function approach. *International Game Theory Review, 10*(3), 303–317.

Pintassilgo, P., Finus, M., Lindroos, M., & Munro, G. (2010). Stability and success of regional fisheries management organizations. *Environmental & Resource Economics, 46*(3), 377–402.

Pintassilgo, P., Kronbak, L. G., & Lindroos, M. (2015). International fisheries agreements: A game theoretical approach. *Environmental & Resource Economics, 62*(4), 689–709.

Pintassilgo, P., Laukkanen, M., Grøenbæk, L., & Lindroos, M. (2018). International fisheries agreements and non-consumptive values. *Fisheries Research, 203,* 46–54.

Punt, M., Weikard, H.-P., & van Ierland, E. (2013). Marine protected areas in the high seas and their impact on international fishing agreements. *Natural Resource Modeling, 26*(2), 164–193.

Sandberg, P. (2010). Rebuilding the stock of Norwegian spring spawning herring: Lessons learned. In *The Economics of Rebuilding Fisheries: Workshop Proceedings* (pp. 219–233). Paris: OECD Publishing.

Schaefer, M. (1954). Some aspects of the dynamics of populations important to the management of commercial marine fisheries. *Bulletin Inter-American Tropical Tuna Commission, 1,* 25–56.

United Nations. (1995). *United Nations conference on straddling fish stocks and highly migratory fish stocks. Agreement for the implementation of the provisions of the United Nations convention on the law of the sea of 10 December 1982 relating to the conservation and management of straddling fish stocks and highly migratory fish stocks.* UN Doc. A/Conf./164/37.

Chapter 7
Other Fishery Game Approaches

Abstract There is a continuum of approaches to model the strategic behaviour between and among agents in fisheries. The chapters so far have introduced the readers to some of the most widely used cooperative and non-cooperative fishery games. The aim of this chapter is to introduce the reader to another set of approaches. The chapter will present the concepts of multi-stage games and repeated games. It will, furthermore, address a class of games often referred to as fish war games, as well as games with spatial dimensions, and games under uncertainty. The chapter concludes with a case study on the groundfish trawl fishery of British Columbia.

7.1 Introduction

This chapter is designed to provide a brief overview of other game approaches to fisheries management, not covered in the previous chapters. It has a different format than the previous chapters, since the intention is to be broader in its context as an overview. The chapter will not present the reader with rigorous formal definitions and detailed examples encountered in the earlier chapters. Rather, the intent is to provide the reader with a basic intuitive understanding of these other game approaches to fisheries management.

When it comes to strategic interaction, we have so far only presented models where the agents are at the same organizational level (e.g. agents are all countries, producer organizations (POs), fishermen or the like). With inspiration from the contract theory, this chapter introduces the reader to multi-stage games and principal–agent theory and discusses the strategic aspect of regulators making decisions before fishermen.

The chapter also presents repeated games, a class of dynamic games in which agents repeat their interactions through time. The so-called "fish war games" are also described. These are discrete time dynamic games based on the Cournot–Nash model.

Two other topics are then presented to the reader. One is the spatial distribution of the stocks, which affects the profitability of the fishing fleets. This topic nicely links to the next one, which is uncertainty arising from economic and the biological dimensions.

© Springer Nature Switzerland AG 2020
L. Grønbæk et al., *Game Theory and Fisheries Management*,
https://doi.org/10.1007/978-3-030-40112-2_7

7.2   Multi-stage Games and Principal–Agent Theory

In fisheries management, decisions are often not made simultaneously, but rather in sequences. This happens, for example, in cases in which regulations on the fishery are set by the authorities, after which the fishermen decide on their harvest level, or in cases in which fishermen first decide on the coalitions to be formed, following which they decide on the effort levels employed in the fishery. The game-theoretical approach to model this class of strategic interaction is to form games in extensive form, a so-called game tree.

A **stage game** is an extensive form of a game where players can choose a new strategy at the beginning of each stage.

Stage games are often used in the context of repeated games, which will be discussed in the next section.

A Stage-Game Model: The Leader–Follower Model
With reference to the original von Stackelberg leader–follower model, where firms in a duopoly have a sequential quantity choice, we present a modified stage-game model, where a regulator sets a regulation for the exploitation of a resource, e.g. a Total Allowable Catch (TAC), and decides on enforcement. The fishermen decide on which coalitions to form and then on their choice of effort levels. The Stackelberg leadership model is a sort of a principal–agent (PA) model. It is not uncommon to have fishery resource management measures in fisheries being implemented at the intra-EEZ (Exclusive Economic Zone) level, with the fishermen in the EEZ then having to make harvesting effort decisions in response to these measures.

The following presents a general framework with relevance to regulated EEZ fisheries, where management measures are decided upon centrally, but implemented on a decentralized level. The model, schematically presented, in Fig. 7.1, consists of three stages. In the first stage, the fishing authority is the player, whereas in the second and third stages the fishermen are the players. In stage 1, the regulator makes enforcement decisions. In stages 2 and 3, the decision-makers are the fishermen. They decide on coalition formation first and then on their effort level. This framework can

A multi-stage game model for fishery management.

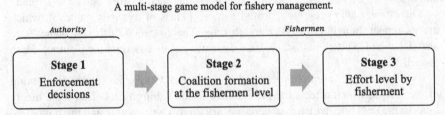

Fig. 7.1 A multi-stage game model for fishery management. *Source:* adapted from Kronbak and Lindroos (2006)

be used to represent fisheries managed by a single government, as it is presented here, but also for fisheries managed by several countries, such as the EU for which the model originally was developed. In this latter case, stage 1 is extended with a stage allowing the countries to form coalitions with respect to their enforcement levels.

The model shows the effects of fishermen forming coalitions. In the first stage of the model, the authority makes decisions and acts as a leader in the von Stackelberg terminology. In the remaining two stages, the fishermen decide on coalition forma-tion and then on effort employed in the fishery. The model is solved by backward induction.

Consider a Gordon–Schaefer model as presented in Chap. 6, Sect. 6.2.1. The additional part to be introduced in the stage model is the authorities control effort decision (stage 1) and the coalition formation decision (the expansion of stage 1). The model is solved backwards, with the fishermen choosing the fishing effort level, for each coalition structure under the level of enforcement. The decision variable for the authorities is the control effort, Z, defined between zero and one, which corresponds to the proportion of fishermen being controlled. The maximization problem for a single intra-EEZ authority is as follows:

$$\max_Z \pi = \sum_{i=1}^{n} P_i(\Psi) - \frac{\gamma}{1-Z}, \quad \text{s.t. } 0 \leq Z < 1, \tag{7.1}$$

where i represents a given fisherman, n is the total number of fishermen and P_i is the fishermen's profit. The probability of being detected is denoted by Ψ, and $\gamma > 0$ is a cost parameter of enforcement effort. The second term on the right-hand side of the equation represents the total costs of control. If the control effort is intensive (e.g. almost full control), Z approaches one, and the costs of control go to infinity. On the other hand, if control effort approaches zero, there are still some fixed management costs.

This is a steady-state model which studies the long-run equilibrium. Hence, the dynamic adjustment towards the equilibrium is not considered. The regulation (in this case a TAC) is exogenous to the model, and the distributional issues are disregarded in this context, i.e. assumed away. The resource manager is a true social planner, who aims to maximize the sum of the profits of the fishermen, net of control costs. Finally, with regard to the fishermen, each player is assumed to attempt to maximize its profits. In so doing, each player will assess the probability of incurring a fine for violating the fishing effort rules. An obvious extension of the model is to set it up as a repeated game, concept which will be discussed at a later point.

The Principal–Agent Theory

As has been presented in the example above, there are some cases where a sequence of choices is made by different players. Often this relates to a principal making a first choice and an agent making a subsequent choice.

A **principal–agent** model is a sequence of decisions made by a principal and an agent often in which asymmetries exist between individuals at the timing of contracting (Mas-Colell et al. 1995).

In some cases, the information asymmetry may be subsequent to the signing of the contract. The contracting parties aim to design a contract that mitigates the difficulties the asymmetric information may cause. The model has several applications in economics, e.g. an owner hiring a manager, insurance companies and their customers, banks and borrowers, regulators and fishermen, and vessel owner and crew. The latter two being the context which we will analyse further. The information asymmetries are often divided into two types of information problems: hidden action/moral hazard and hidden information.

The essence of the PA problem is that the "principal" has a task which he/she/it is unwilling or unable to do him/her/itself and so acquires the services of an "agent". A First-Best situation for the principal would be one in which the actions of the agent could be specified by the principal in a contract and then rigorously enforced. The agent would be no more than a robot.

In a Second-Best situation, the agent has some freedom to choose its actions and is most certainly not a robot. The principal must rest content with establishing an incentive scheme designed to influence the agent's choice of actions. The difference between the economic return to the principal under the Second-Best situation, and what it would have received under a First-Best situation, is the "incentive gap" or "agency cost".

The application of the PA theory to fisheries dates back to Clarke and Munro (1987, 1991), who employ the theory in analysing the issue of coastal states granting access to distant water fishing states (DWFSs) to the coastal state EEZs. They apply the theory using a dynamic resource model. The coastal state acts as principal and the DWFS as agent. It is assumed, as was then reasonable, that the coastal state's incentive scheme consists of a blend of taxes, harvest and effort. It is assumed that we are living in a deterministic world—no uncertainty. Then, even though it is in a world of certainty, a Second-Best outcome for the principal may be inescapable.

Jensen and Vestergaard (2002) presents a PA game using a fisheries steady-state model. The authors combine it with an adverse selection problem, where decisions depend on unobservable characteristics. They assume a socially/centrally owned resource where society collects the revenues that fishermen generate from the resource. Fishermen bear the cost of harvesting and receive a subsidy from society. Since the resource is centrally owned, and the central planner does not have information about which type the fishermen are, the problem directly translates into a principal–agent problem.

Now we come back to the difference between the dynamic resource model and the steady-state resource model as presented in Chap. 3. In the dynamic view of the world, the PA problem arises fundamentally from the inability of the principal to monitor the actions of the agent perfectly. This may arise from uncertainty, the inability of the principal to observe the actions of the agent—information asymmetry, but it may arise, as well, from other factors. In the steady-state view of the world, the PA problem arises from information asymmetry alone.

7.3 Repeated Games

In fisheries, as in many other human activities, the interactions between and among agents are repeated through time. For instance, in the harvest of internationally shared fish stocks, countries usually negotiate every year the total allowable catch. Repeated games are a class of dynamic games designed to capture the long-term interaction between players and explain aspects such as cooperation, threats and punishment (Osborne and Rubinstein 2016).

> A **repeated game** consists of a "base game" or "stage game", which is repeated either a finite or infinite number of times (Maschler et al. 2013).

In a repeated game, the strategy of each player in a given period will depend on the past actions of the other players. This may generate equilibrium outcomes, which would not occur if the game would be played only once (Fudenberg and Tirole 1991). For instance, take the Prisoner's Dilemma game illustrated in Chap. 2, Fig. 2.3, in which the Nash equilibrium is non-cooperation by both players: (NC, NC). If the game is played repeatedly, then one possible strategy is: "cooperate whilst the other player cooperate; if the other player plays non-cooperate then play non-cooperate in every subsequent period". It can be shown that through this strategy the desirable outcome, both players cooperating (C, C), can be supported as a subgame perfect equilibrium of an infinitely repeated game, if the discount rate is sufficiently close to 0, that is, when players give a sufficiently high importance to future payoffs.[1]

According to Osborne and Rubinstein (2016), the key achievement of repeated games is to show the type of strategies that lead to mutually desirable outcomes in a game. These games give insights into the expected structure of behaviour in repeated interactions, which can be termed as social norms. The solution of repeated games shows that, in order to sustain mutually desirable outcomes, the social norms involve the punishment of the players with undesirable behaviour.

A central result of repeated games is that, under certain conditions, all reasonable payoff profiles, and hence also those when all players choose to cooperate, can be

---

[1]A note of caution at this point is appropriate. In a fisheries repeated game, the move towards cooperation can come too late. We have the case of the South Tasman Rise orange roughy fishery, involving Australia and New Zealand, discussed in Chap. 3. In retrospect, this can be seen as a repeated game, with cooperation degenerating into non-cooperation, followed by a subsequent return to cooperation. The return to cooperation came too late, however. By the time that cooperation had been restored, the resource had been driven to near commercial extinction.

achieved as a Nash equilibrium. This result has been widely known in game theory before being published, and hence it has been termed as "folk theorem" (Fudenberg and Tirole 1991).

> **Folk Theorem**: If players are sufficiently patient, then any individual rational payoffs can be a Nash equilibrium of a repeated game.

According to the folk theorem, if players are patient, that is, they have a low discount rate, then desirable outcomes that would not occur, if players were impatient, can be sustained through the use of "trigger strategies". Under these strategies, if a player deviates from a cooperative action, then all the other players will take punitive actions until the end of the game. This sustains cooperation as any finite one-period gain from a deviation is outweighed by even a small loss in every period in the future (Fudenberg and Tirole 1991). Cooperation is thus supported by a social arrangement in which players are deterred from deviating by a threat of punishment (Osborne and Rubinstein 2016). Another implication of the folk theorem is that the set of equilibrium outcomes in a repeated game is large if players are impatient.

Repeated strategic interactions are common in fisheries. Fishermen interact with regulators, countries interact in the context of regional fisheries management organizations, etc. Thus, repeated games can provide important lessons for fisheries management. Hannesson (1997) provide an application of repeated games to fisheries. Using the framework of an infinite horizon repeated game, the author models the harvest of shared fish stocks and shows that cooperative agreements can be supported by trigger strategies.[2] According to these strategies, if a player deviates from the cooperative harvest level then, when detected, all the other players will revert to non-cooperative harvest levels. In the presence of low discount rates, the threat strategies make cooperation an equilibrium outcome.

Grønbæk and Lindroos (2019) present a recent development of a repeated game combining it with the elements of a club good. They present the analytical results for stability issues in a repeated coalition game where information inside the coalition is treated as a club good, which is modelled as a scale parameter on costs. When a player deviates from a coalition, it is excluded from a club good and the games turns into an asymmetric game. The model shows a clear relation between the patience of the players (the discount rate) and the effect of the club good (the scale parameter). Further, the model provides an explanation for the stability of coalitions larger than two players.

---

[2]An earlier application of trigger strategies to fisheries is due to Kaitala and Pohjola (1988). The authors show in a 2-player game on a transboundary fishery resource that trigger strategies coupled with compensatory transfers can sustain cooperation.

7.4 Fish War Games

Conflicts about fishing rights have been analysed through a branch of dynamic games, often called fish war games. It is a branch of literature that extends the discussion of dynamic games presented in Chaps. 3 and 4. This branch of literature evolved based on a dynamic Cournot–Nash model developed by Levhari and Mirman (1980). They introduced a discrete-time non-linear two-player game model with a dynamic externality. Their model has the feature of incorporating both a dynamic and a strategic aspect into in a Cournot–Nash model, and they developed a widely used and cited class of dynamic games using dynamic programming solution principles. The model introduces utility maximizing players acting like Cournot agents in a dynamic duopoly problem. As an add-on, there may be a change in the size of the fish population over time which forms the standard state constraint on the fishery. The model and in particular the objective function is different from the ones presented in Chaps. 3 and 4 since agents are assumed to be utility maximizers. Hence, players derive utility from catches and the cost side of harvesting is implicitly incorporated in the utility.[3] They find, since it is utility that is maximized, that the Cournot–Nash policies imply a greater harvest of fish, and, therefore, a smaller steady state, compared to the models presented in Chaps. 3 and 4.[4]

Fischer and Mirman (1992) present a two-country model with each country fishing and consuming different species of fish, incorporating a biological externality in the interaction between the two species. Assuming that countries only consume one species eliminates the dynamic externality, according to which countries compete for the fish, and allows us to isolate the biological externality, the so-called interspecies interaction. The model is solved for an example and based on the case with negative interaction between species there is less fishing under non-cooperation compared to cooperation.

The two aforementioned models of dynamic Cournot duopoly are combined in the multi-species model by Fischer and Mirman (1996). In their conclusion, they observe that the cooperative case, when both externalities are present, is identical to the cooperative case of Fisher and Mirman's (1992) model with only a biological externality. The main conclusion from their contribution is that non-cooperative behaviour can, in a special case, lead to underutilization of fish stocks due to interdependencies of the stocks. The special case is when there is mutual competition between species meaning the species interact negatively on each other. The reason is that each player does not take into account the effect of its harvesting to the other player's stocks.

Finally, coalition fish war models, with single species, have been recently developed. Kwon (2006) sets a symmetric coalition game in a multiple country context.

---

[3] The dynamic models used in Chaps. 3 and 4 assume that harvesting costs are a function of the size of the biomass (because of the underlying Schaefer model), while the Levhari & Mirman model assumes that harvesting costs are independent of the size of the biomass.

[4] If the harvesting costs are a function of the size of the biomass, we should expect a larger steady-state biomass than would be the case, if such costs are independent size of the biomass.

He concludes that any coalition with more than two member countries cannot be sustained in a game where countries move simultaneously. Furthermore, if the coalition is a dominant player, larger coalitions form and their size depends on the parameters of the problem. Breton and Keoula (2014) present an asymmetric game with many countries which differ in their time preferences and discount rates, belonging to one of two groups: high or low discount rate. They show that impatient players (higher discount rates) have stronger bargaining power. The authors conclude that cooperation can include many countries in the case where the cooperative coalition has a first-mover advantage. As has been seen also in Chap. 6, where a steady-state model is presented, they conclude that asymmetry has a significant impact on the way the resource is shared and on the profitability of coalitions.

7.5 Games with Spatial Dimensions

The spatial distribution of fish stocks is a key element in fisheries. It affects the profitability of the fleets, as harvesting costs increase with the distance to the fishing grounds. It has also implications for access to the resources.

One can think of three spatial areas. There is first the area consisting of the Exclusive Economic Zones (EEZs). If a fish stock is confined to an EEZ, the coastal state has full legal right to restrict access to the stock. A second area consists of high seas, subject to no regulation. Fish stocks in this area are common-pool resources and can be accessed by any fleet. In between are high seas under the jurisdiction of Regional Fisheries Management Organizations (RFMOs). As discussed in Chap. 1, in legal terms, the control over access in RFMOs, while not great as it is in EEZ, is significant.[5] Having said this, RFMOs differ considerably in their governance ability. Some are successful in suppressing unregulated fishing—free riding; others are most decidedly not. Chapter 6 provides with two case studies, one of a successful RFMO type of arrangement, and one of an RFMO that has been remarkably unsuccessful.

Finus et al. (2020) provide an example of a fishery game with spatial dimension. They model an internationally shared fishery by incorporating the spatial dimension of the stock within a partition function game (presented in Chap. 6). It is assumed that the stock is distributed over a given area, which is divided into high seas and EEZs. The authors simplify, by assuming that the high seas under investigation is subject to no effective governance. Their high seas can be seen to consist of high seas beyond RFMO jurisdiction, along with those under the jurisdiction of weak RFMOs.

The authors explore how the characteristics of excludability and rivalry vary in different settings. Two forms of excludability are considered. *Technical excludability* is related to intrinsic properties of the stock, such as the migration between different zones. A more intense migration lowers *technical excludability*. *Socially constructed excludability* is set by society, such as the establishment of EEZs. The larger the EEZs, the higher is the *socially constructed excludability*, as coastal states can exclude the

---

[5]See Chap. 1.3.1.

other fishing states from harvesting in their EEZs. The authors associate *rivalry* with the intrinsic growth rate of the stock. The higher this rate be, the lower the *rivalry* between states.

In this setting, Finus et al. (2011) reach interesting results for a game with three symmetric players. First, under full non-cooperation, in which the three players act as singletons, the aggregate fishing effort (sum of the effort of the three players) increases with the share of the high seas in the total fishing area and with the intensity of stock migration. That is, the lower the *socially constructed* and the *technical excludability,* the higher will be the aggregate fishing effort. This translates into lower fish stock and lower individual and total payoffs (profits). Hence, a lower portion of the stock share in the high seas contributes to the preservation of the stock and to higher economic rents.

Second, the aggregate fishing effort decreases with the level of cooperation between states. The grand coalition, formed by the three players, yields the lowest aggregate fishing effort, when compared to partial cooperation, between two states, and to full non-cooperation. Thus, more cooperation reduces the aggregate fishing effort and increases the stock level. It has also a positive impact on the total payoffs.

Third, it is shown that the grand coalition is never an equilibrium outcome of the game, due to the presence of strong free-rider incentives. Moreover, partial cooperation is only stable in the presence of high excludability (low high seas portion and low intensity of stock migration). It turns out that high excludability is the setting in which cooperation, both full and partial, leads to the lowest increase in total payoffs, compared to non-cooperation. That is, cooperation has the lowest impact. Conversely, when cooperation has a large impact, the equilibrium of the game is full non-cooperation. Thus, it is concluded that the "paradox of cooperation" (Barrett 1994) also applies to international fisheries: cooperation is weak when it is most needed.

Liu and Heino (2013) also address the impact of the spatial distribution of a fish stock in international fisheries. The authors adopt a dynamic non-cooperative game with two players, in which the distribution of the stock shifts due, for instance, to climate change. They compare the outcomes of reactive management, in which the manager ignores future shifts, with proactive management, in which such shifts are considered. The results show that under reactive management the two players are symmetric and hence present the same strategies. Under proactive management, players anticipate future shifts in the distribution of the stock, which will affect them differently. It is shown that the player losing the stock tends to harvest more aggressively than the player gaining the stock.

Diekert and Nieminen (2017) also analyse international fisheries using a two-player dynamic non-cooperative game with spatial dimension. In their model, the stock shifts its spatial distribution at some uncertain moment in time, which originates a change in ownership from one state to the other. It is shown that the slower the shift (larger transition period) the lower will be the total extraction rate (ratio of harvests to the stock) and the better the chances to achieve cooperation between the two states. Sudden shifts in the spatial distribution of a fish stock lead to aggressive exploitation of the resource and make cooperation unlikely.

7.6   Games Under Uncertainty

Fisheries are usually characterized by significant uncertainties arising from biological and economic dimensions. The former is due to environmental causes and can affect stock recruitment, growth, mortality, spatial distribution and the carrying capacity of the environment. The latter is related to fish prices, harvesting costs, technology and fishing regulations. Despite the relevance of uncertainty in fisheries, the vast majority of the applications of game theory to this activity have adopted a deterministic setting. Herein we highlight some of the exceptions.

An early application of game theory to fisheries in a context of uncertainty is due to Kaitala (1993). The author considers stock uncertainty in a two-player dynamic game, in which countries choose between cooperation and non-cooperation in every period. The choice between these two actions, for each player, depends on the fishing effort of the other player in the previous period, regarding which there is imperfect information. It is shown that, in the equilibrium, periods of cooperation can alternate with periods of non-cooperation. This feature, observed in many internationally shared fisheries worldwide, stems from imperfect information.[6]

Another example of a game with stock uncertainty and imperfect information is due to Laukkanen (2003). The study incorporates uncertainty through recruitment shocks, in a two-player dynamic game. The model is applied to the Baltic salmon fishery, in which the harvest is undertaken sequentially by two fleets: commercial offshore and inshore. The harvests undertaken by the inshore fleet are not observed by the offshore fleet. The results show that stochastic shocks in recruitment may trigger phases of non-cooperation. Moreover, the larger the uncertainty, the less likely is cooperation. Laukkanen (2005) extends the analysis to implementation uncertainty: the variation between management targets and management outcomes. The author concludes that implementation uncertainty is more likely to hamper cooperation than recruitment uncertainty.

Miller et al. (2013) discuss the effect of stock uncertainty due to climate change on the governance of internationally shared fish stocks. The authors stress that imperfect information coupled with environment changes makes it difficult to assess the contribution of any country to a cooperative agreement, which may hamper cooperative management agreements.

Another study on stock uncertainty is due to Miller and Nkuiya (2016), who consider uncertainty on stock growth, in the form of a threat of regime shift. The authors use a two-stage game to model coalition formation on the management of the stock. They show that uncertainty affects not only harvest decisions, but also the incentives for cooperation and hence the number of countries joining the fisheries agreement (coalition size). Among other results, the authors show that an exogenous shock of total stock collapse may destabilize coalitions and increase total harvests. They also show that for particular forms of uncertainty a counterintuitive result may arise: uncertainty can foster cooperation, by increasing coalition size. This result

---

[6]An example of cooperation alternating with non-cooperation, due to imperfect information, is provided by the case study on Pacific salmon off Canada and the United States, presented in Chap. 4.

has also been found in the literature on international environmental agreements, for very specific forms of uncertainty (e.g. Kolstad and Ulph 2011). The basic idea is that under a "veil of uncertainty" free-rider incentives may decrease and hence larger coalitions may form. Although uncertainty is generally an impediment to cooperation, this study shows that, in special cases, the reverse can be true.

A different approach is taken by Mina et al. (2016), who address the behaviour of fishermen under uncertainty. The authors undertook an experimental study on Mexican and Colombian fishermen regarding the effects of climate change on fisheries. It is concluded that fishermen show risk aversion and are ready to reduce harvests and change fishing practices if they are provided with sufficient information on the consequences of climate change. They also manifest social preferences in aspects such as collaboration and reciprocity, which can play an important role in achieving cooperative management of the fish stocks.

Punt (2018) incorporates uncertainty on the fishing costs in a transboundary fishery. This type of uncertainty emerges from the relocation of fish stocks due to climate change, which may attract new entrants. A multi-stage game with two players is adopted in a context in which the entry into the fishery requires sunk investment costs. It is concluded that the sunk costs can increase the competitive pressure on the fish stock and act as an entry deterrence mechanism and a commitment device for new entrants.

7.7 Case Study: The Groundfish Trawl Fishery of British Columbia, Canada

This case study, of an intra-EEZ fishery, will serve to illustrate multi-stage games, repeated games and uncertainty. Furthermore, it will point to issues to be raised in the following chapter, namely, evolutionary games and multi-level cooperative games.

The British Columbia (B.C.) groundfish trawl fishery is an extraordinarily complex multi-species fishery involving over 60 stocks extending throughout the length and breadth of the British Columbia 27,000 km. coastline, involving species such as various types of rock cod, Pacific cod, hake and lingcod (Munro et al. 2009; Wallace et al. 2015).[7] When Canada implemented its EEZ regime in 1977, it eliminated foreign groundfish vessels from its Pacific EEZ and encouraged the expansion of the Canadian Pacific groundfish fleet. The first formal management plan for the fishery was introduced in 1980, consisting of TACs for the relevant sub-fisheries and a limited entry programme, in which vessels granted access to the fishery were allowed, indeed encouraged, to compete for shares of the TACs, popularly referred to as Olympic style TACs.

---

[7]It is claimed that, if British Columbia were an independent country, it would have the eight longest coastline in the world (Munro 2017).

This can be viewed within the context of the Kronbak-Lindroos multi-stage game, discussed in Sect. 7.2. The resource manager took the form of the federal Department of Fisheries and Oceans (DFO).[8] DFO could have been seen as playing a von Stackelberg leader–follower game with the fishermen—the industry. The coalition formation of the fishermen was that of full competition—every fisherman player a singleton. The competitive fisher game, combined with the competitive fishermen–resource manager game, gave powerful testimony to the predictive power of the Prisoner's Dilemma. By the mid-1990s, the resource manager was forced to concede that it had a management disaster on its hands. Overcapitalization of the fishing fleet was manifest, which in turn effectively undermined the resource manager's ability to control fishing effort and catches (Munro et al. 2009).[9] In 1995, DFO took the unprecedented step of shutting the fishery down.

Over the years 1996–1997, DFO re-opened the fishery by introducing step-by-step major reforms. Surveillance and enforcement of the fishery were greatly enhanced, with the cost of the enhanced surveillance and enforcement to be borne entirely by the industry. While the TACs and limited entry were retained, the Olympic style nature of the TACs was replaced by an ITQ scheme, in which vessel owners were given portfolios of quotas to the relevant species and then invited to become portfolio managers. Think of a new stage game, in which the resource manager comes forth with a new set of management measures in stage 1. The question then became: how would the fishermen respond in stage 2?

The received wisdom among fishery economists at that time was that the scope for achieving a stable cooperative game among ITQ holders was very limited, with the number of players $n = 15$ seen as the magic upper limit (Townsend 2010). In addition, the prospects of an ITQ scheme having any degree of success in a multi-species fishery were seen as bleak, since multi-species fisheries are simply far too complex (Squires et al. 1998).

The vessel ownership relations in the B.C. groundfish trawl fishery were (and are) complex, but it has been argued that the number of independent players is: $n \geq 30$ (Munro 2017). The fishery is nothing, if not multi-species in nature. According to the received wisdom, the ITQ scheme should never have worked, but work it did.

Think of the fishermen game as being, not a "one-shot" game, but rather a repeated game, with evolutionary aspects.[10] The consequence of this repeated fishermen game, in spite of the received wisdom, was the emergence of a stable grand coalition. Clear evidence of this state of affairs emerged by the year 2000. Early in that year, fishermen came to suspect that a valuable resource in the mix, Pacific cod, was in significant decline (Munro 2017).

---

[8]In contrast to some other federal coastal states, the management of Canada's intra-EEZ marine fisheries resources is the sole prerogative of the Canadian federal government.

[9]To take but one example, a valuable species in the fishery is Pacific Ocean Perch. By the mid-1990s, the actual harvests for the sub-fishery, based on this resource, were exceeding the TAC by 100%.

[10]To be discussed in Chap. 8.

The fishing industry has an industry-wide organization, which facilitates communication among the players, which has no control over individual players.[11] Through that organization, the fishermen urged DFO to reduce drastically the TAC on Pacific cod and to carry out further stock assessment. DFO did the latter. The further stock assessment revealed that the industry's fears were fully justified. The industry then recommended a sharply reduced TAC for the following year. DFO accepted the recommendation.

What then followed was an industry lead investment programme in the resource, an investment programme that had a strongly positive payoff (Munro 2017). In 2006, there was a similar case, involving another valuable resource, Pacific Ocean Perch. In this case, the industry not only pressured DFO into reducing the TAC on the resource for several years, but also set out to hire at industry expense a scientist to work with DFO scientists on Pacific Ocean Perch stock assessment.[12]

In addition to all of this, the industry, commencing in the mid-2000s, negotiated a habitat agreement with a consortium of NGOs. The NGOs had attacked the industry for habitat destruction through bottom trawling, with a set of sponge and coral species being of specific concern. The attacks had economic teeth, because they had the power to create barriers for the industry in the important California market.[13]

Due to a legal anomaly, the resource manager could do nothing. The industry had to take the initiative, which it did with the full support of DFO. An agreement was eventually negotiated with the NGO consortium, which was then incorporated into the DFO management plan in early 2012.[14] Under the terms of the Agreement, the groundfish fleet is given a small annual global sponge/coral global quota, spread out among the vessels like tiny ITQs. Any vessel that exceeds its quota must cease fishing for the season. The seventh year of the Agreement concluded in February 2019. In all seven years, the actual catch of sponge and coral fell far below the allowed quota.[15]

A question and several comments follow from all of this. The question is: the fishermen game has exhibited "internal stability"—why? What has prevented massive free riding? The response is that there is as of yet no clear answer. Further research is required.

There is, however, a two-part conjecture to explain the stability of the fishermen game. First, there is reason to believe in the existence of club goods.[16] Secondly, in the discussion in Chap. 6 of RFMOs, the point was made that RFMO cooperative

---

[11] The Canadian Groundfish Research and Conservation Society.

[12] Through the industry organization, it was agreed to establish a research fund, with a formula worked out for vessel owner contributions. The contributions are strictly voluntary.

[13] See Clark and Munro (2017) and Munro (2017) for details.

[14] B.C. Groundfish Trawl Habitat Conservation Collaboration Agreement (Wallace et al. 2015).

[15] For administrative purposes, the groundfish trawl season ends on February 20, with the new season beginning on the 21. The annual global sponge/coral quota is 4500 kg. In the 2018/2019 season, the actual sponge/coral catch was 340 kg (Turris 2019).

[16] To take one small example, "club goods" can be seen to include sanctions and peer pressure. It was noted that the groundfish trawl fishery industry has established a research fund to which vessel owners contribute annually on a voluntary basis. What ensures compliance; prevents free riding? Peer pressure is the answer given (Bruce Turris personal communication).

games have to be self-enforcing, as there is no third party to be turned to for assistance in curbing free riding. In the case of intra-EEZ fishermen cooperative games, there is a potential third party, namely, the domestic resource manager. The thorough surveillance and enforcement programme imposed upon the industry by the resource manager, DFO, must certainly act to suppress free riding (Munro 2017).

The first comment is that in the mid-1990s, the Kronbak–Lindroos leader–follower game seemed to fit this fishery perfectly. At the time of writing, that leader–follower game is no longer valid for the fishery. To begin, a leader–follower game is inherently non-cooperative. Through a process of evolution, which at this time is not at all fully understood, the non-cooperative resource manager–fishermen game has been transformed into a cooperative resource manager–fishermen game.[17] It has been argued that the management of the fishery can be characterized as de facto co-management (Munro 2017).[18] Second, it has been seen that on several occasions, the leader–follower role has been reversed, with the industry, the fishermen collectively, being the leader, and DFO the follower.

The second comment pertains to uncertainty. The de facto co-management of the resource, the cooperative fishermen–resource manager game, which we can term multi-level cooperation, can be seen to mitigate the consequences of uncertainty. In Principal–Agent analysis, it is common to talk about information asymmetry, where the regulated have more information than the regulator. So to with the fishery described, where it can be argued that the fishermen often have more information than the resource manager. With the multi-level cooperation described, this asymmetry can be seen to work in favour of the resource manager.

The Pacific cod and Pacific Ocean Perch cases discussed involved uncertainty, uncertainty with respect to the state of the resources. Asymmetry of information arose in these cases, because the industry recognized before the resource manager that the resources were in decline. The industry, in effect, acted as an early warning system, to the clear benefit of the resource manager and of the ultimate owners of the resources, the Canadian public.

7.8   Policy Implications

This chapter shows that the type of game to be applied depends significantly on the context of the problem. As a result, the different games provide different sets of policy implications.

The policy implications of the multi-stage game relate to the discussion of regulations. When new regulations are implemented, they have a spillover effect on

---

[17]The evolution can be seen to have commenced with the Pacific cod case in 2000. This was a clear example of cooperation between the fishermen and the resource manager. Successful cooperation led to further cooperation—success breeds success.

[18]The de facto aspect has to be stressed. The industry can influence the management of the fishery, but cannot determine it. The final resource management decisions rest with the Minister of Fisheries and Oceans (Munro 2017).

the behaviour of those who are regulated. Change in regulations may create incentives to a different behaviour, sometimes to the good sometimes in perverse or unintended directions. There is a large potential for applying multi-stage games to predict behavioural changes by the fishermen and thereby analyse the consequences of regulation on the exploitation of a resource. For instance, they can be applied to address intra-EEZ regulation, for instance, to the case study presented in this chapter: The groundfish trawl fishery in British Columbia.

The repeated games show that repeated interaction between agents through time can lead to mutually desirable outcomes if threats of punishment are used. Thus, when players repeat their interaction, cooperation becomes a more likely outcome. Also using a dynamic framework, fish war games highlight the severe economic consequences of conflicts between harvesting fleets.

Incorporating the spatial dimension of fish stocks in game-theoretical analysis provides important insights. First, the larger the share of the fishing grounds located in the high seas, where the resources can be considered as common property, the more intense will be stock exploitation and lower the payoffs for the fishing states. Second, more cooperation tends to improve the status of the stocks and increase the aggregate payoffs. Third, cooperation is more likely in fisheries characterized by high excludability (e.g. where a large proportion of the fishing area is within EEZs). However, in these cases, the gains of a cooperative agreement compared to non-cooperation tend to be relatively low. The "paradox of cooperation" applies. Fourth, the type of expectations that players have about future shifts in the distribution of stocks plays a key role in their strategic interaction. Players anticipating a future loss of fishing grounds will react with aggressive harvesting strategies. Fifth, the duration of the transition periods caused by distributional shifts impacts not only on the harvesting strategies but also on the likelihood of cooperation. In particular, larger transitional periods lead to lower harvest rates and are conducive to cooperation.

A few key lessons can be taken from the applications of game theory to fisheries under uncertainty. First, uncertainty together with imperfect information may generate non-cooperation phases, which alternate with cooperation phases. Second, the larger the uncertainty the less likely is to achieve cooperation. Third, implementation uncertainty may be more decisive in impairing cooperation than stock uncertainty. Fourth, uncertainty impacts not only on the fishing strategies of each country but also on the formation of international fisheries agreements. Fifth, fishermen are risk averse and hence, if properly informed and coordinated, may be ready to cooperate in reducing the impacts of uncertainty, namely, those accruing from climate change. Finally, uncertainty on the spatial distribution of stocks, caused, for instance, by climate change, may attract new entrants to a given fishery making previous agreements unstable.

References

Barrett, S. (1994). Self-enforcing international environmental agreements. *Oxford Economic Papers, 46,* 878–894.

Breton, M., & Keoula, M. Y. (2014). A great fish war model with asymmetric players. *Ecological Economics, 97,* 209–223.

Clark, C., & Munro, G. (2017). Capital theory and the economics of fisheries: Policy implications. *Marine Resource Economics, 32,* 123–142.

Clarke, F. H., & Munro, G. R. (1987). Coastal states, distant water fishing nations and extended jurisdiction: A principal–agent analysis. *Natural Resource Modeling, 2*(1), 81–107.

Clarke, F. H., & Munro, G. R. (1991). Coastal states and distant water fishing nations: Conflicting views of the future. *Natural Resource Modeling, 5*(3), 345–369.

Fudenberg, D., & Tirole, J. (1991). *Game theory.* Cambridge: The MIT Press.

Diekert, F., & Nieminen, E. (2017). International fisheries agreements with a shifting stock. *Dynamic Games and Applications, 7*(2), 185–211.

Finus, M., Schneider, R., & Pintassilgo, P. (2020). The role of social and technical excludability for the success of impure public good and common pool agreements: The case of international fisheries. *Resource and Energy Economics, Vol. 59 Online first.*

Fischer, R. D., & Mirman, L. J. (1992). Strategic dynamic interaction: fish wars. *Journal of Economic Dynamics and Control, 16*(2), 267–287.

Fischer, R. D., & Mirman, L. J. (1996). The compleat fish wars: Biological and dynamic interactions. *Journal of Environmental Economics and Management, 30*(1), 34–42.

Gibbons, R. (1992). *A primer in game theory.* Harlow: Pearson Education Ltd.

Grønbæk, L., & Lindroos, M. (2019) *Cooperation and club goods: Fisheries management in the spirit of Elinor Ostrom.* Working Paper submitted for publication.

Hannesson, R. (1997). Fishing as a supergame. *Journal of Environmental Economics and Management, 32,* 309–322.

Holden, M., & Garrod, D. (1996). *The common fisheries policy: Origin, evaluation and future.* Oxford: Fishing News Books Ltd.

Jensen, F., & Vestergaard, N. (2002). A principal-agent analysis of fisheries. *Journal of Institutional and Theoretical Economics, 158,* 276–285.

Kaitala, V. (1993). Equilibria in a stochastic resource management game under imperfect information. *European Journal of Operational Research, 71,* 439–453.

Kaitala, V., & Pohjola, M. (1988). Optimal recovery of a shared resource stock: A differential game model with efficient memory equilibria. *Natural Resource Modeling, 3*(1), 91–119.

Kolstad, C., & Ulph, A. (2011). Uncertainty, learning and heterogeneity in international environmental agreements. *Environmental & Resource Economics, 50,* 389–403.

Kronbak, L. G., & Lindroos, M. (2006). An enforcement-coalition model: Fishermen and authorities forming coalitions. *Environmental & Resource Economics, 35*(3), 169–194.

Kwon, O. S. (2006). Partial international coordination in the great fish war. *Environmental & Resource Economics, 33*(4), 463–483.

Laukkanen, M. (2003). Cooperative and non-cooperative harvesting in a stochastic sequential fishery. *Journal of Environmental Economics and Management, 45,* 454–473.

Laukkanen, M. (2005). Cooperation in a stochastic transboundary fishery: The effects of implementation uncertainty versus recruitment uncertainty. *Environmental & Resource Economics, 32,* 389–405.

Levhari, D., & Mirman, L. J. (1980). The great fish war: An example using a dynamic Cournot-Nash solution. *The Bell Journal of Economics, 11*(1), 322–334.

Liu, X., & Heino, M. (2013). Comparing proactive and reactive management: Managing a transboundary fish stock under changing environment. *Natural Resource Modeling, 26*(4), 480–504.

Maschler, M., Solan, E., & Zamir, S. (2013). *Game theory.* Cambridge: Cambridge University Press.

Mas-Colell, A., Whinston, M. D., & Green, J. R. (1995). *Microeconomic theory*. Oxford University Press.

Miller, S., & Nkuiya, B. (2016). Coalition formation in fisheries with potential regime shift. *Journal of Environmental Economics and Management, 79*, 189–207.

Miller, K., Munro, G., Sumaila, R., & Cheung, W. (2013). Governing marine fisheries in a changing climate: A game-theoretic perspective. *Canadian Journal of Agricultural Economics, 62*, 309–334.

Mina, J., Fernández, D., Ibarra, A., & Georgantzis, N. (2016). Economic behavior of fishers under climate-related uncertainty: Results from field experiments in Mexico and Colombia. *Fisheries Research, 183*, 304–317.

Munro, G. (2017). Escaping the subsidies trap: The role of cooperative fisheries management. *Proceedings of the North American Association of Fisheries Economists Forum 2017*, La Paz, Mexico, March 2017, Corvallis, NAAFE.

Munro, G., Turris, B., Clark, C., Sumaila, U. R., & Bailey, M. (2009). *Impacts of harvesting rights in Canadian pacific fisheries* (pp. 1–3). Ottawa: Fisheries and Oceans Canada, Statistical and Economic Analysis Series Publication No.

Osborne, M. J., & Rubinstein, A. (2016). *A course in game theory*. Cambridge, MA: Phi—MIT Press.

Punt, M. (2018). Sunk costs equal sunk boats? The effect of entry costs in a transboundary sequential fishery. *Fisheries Research, 203*, 55–62.

Squires, D., Campbell, H., Cunningham, S., Dewees, S., Grafton, R. Q., Herrick, S., et al. (1998). Individual transferable quotas in multispecies fisheries. *Marine Policy, 22*, 135–159.

Townsend, R. (2010). Corporate governance of jointly owned fisheries rights. In R. Q. Grafton, R. Hilborn, D. Squires, M. Tait, & M. Williams (Eds.), *Handbook of marine fisheries conservation and management* (pp. 520–531). Oxford: Oxford University Press.

Turris, B. (2019). Fisheries management and theory operationalized. Presentation (ppt.) at the Vancouver School of Economics, University of British Columbia, March 2019.

Wallace, S., Turris, B., Driscoll, J., Bodtker, K., Mose, B., & Munro, G. (2015). Canada's pacific groundfish trawl habitat agreement: A global first in an ecosystem approach to bottom trawl impacts. *Marine Policy, 60*, 240–248.

Chapter 8
Conclusions

Abstract As the concluding chapter of the book, this chapter has two purposes. The first is to provide an overview of the main conclusions, the primary lessons, arising from the previous seven chapters. The second is to point to major gaps in the literature, pertaining to game theory and fisheries management, and thus to look forward to the future challenges in the application of game theory to fisheries management.

8.1 Major Lessons and Conclusions

This section points the reader to the major lessons and conclusions arising from the book thus far. These lessons and conclusions are discussed, not on a chapter by chapter basis, but rather in terms of the classical branches of game theory: non-cooperative and cooperative games.

8.1.1 Non-cooperative Games

A non-cooperative game is one in which commitments (agreements, promises, threats) are not enforceable, and hence each player acts independently of the others (see Sect. 2.3). The theory of non-cooperative games is overarching in the discussion and analysis of game theory and fisheries management. There is no chapter in which the theory of non-cooperative games is entirely absent. In Chap. 1, it is pointed out that, while one can with ease develop non-cooperative game models, in which there is no hint of cooperation, the converse is decidedly not true. Behind every cooperative game, it is stated in that chapter, there lurks a non-cooperative game. Chapters 2–7 serve to verify this assertion.

Prior to reviewing the details of the theory of non-cooperative games, we can put forward the policy implications of non-cooperative games for fisheries management up front and succinctly. In Chap. 1, it is stated that non-cooperative management of a transboundary fishery resource will, except in unusual circumstances, result in inferior resource management. This statement we would now, without hesitation,

© Springer Nature Switzerland AG 2020

L. Grønbæk et al., *Game Theory and Fisheries Management*,

https://doi.org/10.1007/978-3-030-40112-2_8

apply to all other internationally shared fishery resources and to many domestic fishery resources. Case study after case study in the following chapters validates this statement. One need only think of the South Tasman rise orange roughy fishery and North Sea herring in the pre-EEZ regime era. Where strategic interaction is involved in the management of a fishery resource, cooperation does indeed matter.

Chapter 2, by way of introduction, sets forth one of the most famous of all non-cooperative games: the Prisoner's Dilemma, which has a wide application in fisheries, where it is often termed the Fisher's Dilemma. The payoffs of this game are such that non-cooperation is the dominant strategy for all players. That is to say, regardless of the actions of the other players, the best response is always non-cooperation. Hence, in the equilibrium of the game, all players behave non-cooperatively.

The dilemma is that all players could receive a higher payoff, if they could agree to cooperate. The strategic interaction between players, however, leads to the trap of non-cooperation. In terms of fisheries, the Fisher's Dilemma brings with it assurances of inferior resource management.

With introductions in hand, Chap. 3 introduces two classical non-cooperative fishery games: one static and another dynamic. The static game is based on the Gordon–Schaefer bioeconomic model, while the dynamic game discussed owes it origins to Clark (1980). Symmetric and asymmetric games are explored.

The main difference between the static and dynamic non-cooperative games as presented in Chap. 3 lies in their predictions with respect to the severity of the consequences of non-cooperation. In the case of static symmetric non-cooperative games, the management of the fishery resource will normally be sub-optimal, but the consequences are not all that bad. In order for non-cooperation to lead to Bionomic Equilibrium, the number of players, n, has to be $= \infty$. In the case of the dynamic symmetric non-cooperative fishery game, non-cooperation will lead to Bionomic Equilibrium, in which the resource rent is fully dissipated, even if n does not exceed 2.

With respect to asymmetric non-cooperative fisher games, both the static and dynamic versions predict an equilibrium biomass level above the Bionomic Equilibrium level, but once again the dynamic model predicts a more drastic outcome. In the static game, the most efficient player will play a larger role in the fishery than its less efficient competitors, and may even drive out some of the less efficient competitors, but this is unusual. In the dynamic game, the more efficient player will drive out *all* less efficient competitors. This can be seen as a type of entrance deterrence strategy, in which the most efficient player keeps the biomass at a level sufficiently low to block the entry of less efficient competitors.[1]

Dynamic games are usually seen as superior to static ones, for the simple reason that the latter are constrained to say the least by their underlying assumptions. That being said, the static game is far, far simpler to apply than the dynamic one.

---

[1] These seemingly sharp differences have to be qualified to a certain degree. In the case of the static symmetric non-cooperative fishery games, for some parameter values, one can in fact approach a situation, which is close to that of Bionomic Equilibrium, with n far below ∞. With respect to the dynamic non-cooperative fishery games discussed in Chap. 3, many simplifying assumptions are used. More complex dynamic non-cooperative fishery games will be found to yield less drastic results.

When combining non-cooperative games with coalition formation, as is done in Chaps. 5 and 6, dynamic non-cooperative games are simply too complex. Static non-cooperative games enable us to make tractable the intractable. Moreover, it can be argued that static non-cooperative games are appropriate when the adjustments to the long-run equilibria are rapid.

Beyond that, there is the question of the predictive power of the static and dynamic non-cooperative fishery models. There is, in fact, not a clear-cut winner. Among the case studies in the book, one pertains to the anchovy resource shared by Peru and Chile and managed by the two non-cooperatively. The outcome has a definite Fisher's Dilemma flavour to it, but there is no evidence that the resource is in grave danger—a static non-cooperative game type of situation.

On the other hand, the book has as another case study, the South Tasman Rise orange roughy fishery resource shared by Australia and New Zealand. A breakdown in the cooperative management of the resource in the late 1990s resulted in near commercial extinction of the resource, a state of affairs, which persists to the time of writing. The book has as well case studies on North Sea herring and Norwegian Spring Spawning herring. In both cases, non-cooperative management threatened the resources with extinction. All three cases demonstrate that the dynamic non-cooperative fishery game model is not without its predictive power.

In moving forward to Chap. 7, this chapter outlines, but does not discuss in detail, five other types of non-cooperative games that have been applied to fisheries. The first consists of multi-stage games, in which players take decisions in different stages. This includes the von Stackelberg leader–follower game, and the principal–agent game. These games make it possible to incorporate, among other matters, the sequence in which the different fishery actors take decisions, as well as asymmetric information. Second on our list consists of repeated games, in which there is a base game that is repeated over time. This captures the fact that in fisheries the interactions between agents are commonly repeated over time (e.g. in an internationally shared fishery countries may negotiate a total allowable catch every year). In a repeated game, if the players have low discount rates, then cooperation can be sustained through the use of "trigger strategies". This makes it possible to break free from the trap of non-cooperation, which cannot be done in a "one-shot" Prisoner's (Fisher's) Dilemma game.

Fish wars games, third type on our list, consist of dynamic games, based on the Cournot–Nash model developed by Levhari and Mirman (1980). This model is different from the classical dynamic game by Clark (1980) discussed in Chap. 3, which rests upon the Schaefer model, leading to the consequence that harvesting costs are a function of the size of the biomass. In the Levhari and Mirman model, such costs are independent of the size of the biomass.

The original Levhari and Mirman model has been extended to capture aspects such as multi-species and coalition formation. One of the reasons for the success of Levhari and Mirman (1980) model, as opposed to the Clark (1980) model, lies in the simplicity of its solution.

Games with spatial dimension, the fourth on our list, have shown important lessons regarding the importance of this dimension on the strategic interaction between players. For instance, cooperation is more likely in fisheries where a large proportion of the fishing area is within EEZs. Another result forthcoming from these games is that players anticipating a future loss of fishing grounds will react with aggressive harvesting strategies.

Fifth in our list consists of games under uncertainty, which have been applied increasingly to fisheries. It has been shown that uncertainty may generate non-cooperation phases. Uncertainty impacts not only on the fishing strategies but also on the formation of international fisheries agreements.

8.1.2 Theory of Cooperative Games

A fundamental conclusion arising from of the analysis of non-cooperative games applied to fisheries is that non-cooperation leads to inferior, sometimes disastrously inferior, resource management. Cooperation does indeed matter, which means in turn that we must examine cooperative fisher games, and do so with care.

A cooperative game is one in which commitments are *fully* binding and enforceable. This is, to say the very least, a demanding requirement. The challenge posed by cooperation, by cooperative games applied to fisheries, is to ensure that the aforementioned commitments remain fully binding and enforceable through time.[2] If the commitments do not do so remain, the cooperation will break down. The cooperative game will degenerate into a non-cooperative game, with all that that implies.

Cooperative games are complex, much more so than non-cooperative games. To begin, the number of players in a cooperative game matters, and matters a great deal. In non-cooperative games, there is not that much difference between an $n = 2$ game and an $n > 2$ game. The latter can be seen as a generalization of the former.[3] In cooperative games, the difference in degree between an $n = 2$ game and an $n > 2$ game is so great as to constitute a difference in kind. With $n > 2$ games, one is plunged into the world of coalitional games.

If the set of players is given by $N = \{1, 2, \ldots, n\}$, then the number of possible coalitions is equal to 2^n, including the Grand Coalition (all players), sub-coalitions, the players as singletons and the empty coalition, $\emptyset$. If $n = 2$, there are four possible coalitions, but only one that really counts—there are no sub-coalitions. If $n = 3$, there are 8 possible coalitions, with 3 possible sub-coalitions. If $n = 7$, not a particularly large game, there are 128 possible coalitions, with 119 possible sub-coalitions.

Cooperative games are introduced briefly in Chap. 2. The first chapter to discuss cooperative games at length is Chap. 4. The chapter is devoted exclusively to $n = 2$ cooperative games, because of their relative simplicity. In the chapter, the basic

---

[2]The cooperative games that are examined are, without exception, dynamic. Static cooperative games do not add to our store of knowledge.

[3]As Chap. 3 clearly demonstrates.

conditions that must be met for the solution to a cooperative game to be stable through time, the individual rationality and collective rationality conditions, are discussed. The role of side payments, referred to in later chapters as transferable utility, in achieving these conditions is examined. What is alluded to, but not discussed in detail, is the need for these cooperative games to be able to withstand unpredictable shocks through time—to be resilient.

Two-player cooperative fisher games have proven to be valuable in analysing the economic management of transboundary stocks, where the number of players is typically small. Such games, however, have proven to be wholly inadequate for analysing the economic management of straddling stocks under RFMO management, where the number of players is large, often very large. Cooperative games with $n > 2$ players must be brought to bear.

With $n > 2$ players in a cooperative game, there are two fundamental questions that have to be addressed: (i) the stability issue—which coalitions will succeed in forming binding agreements? (ii) If a coalition is established, how will the benefits from cooperation be shared among the coalition members? These two questions are, of course, linked. If the sharing of the benefits from cooperation is deemed to be manifestly unfair, this will obviously serve to undermine all attempts to achieve stability.

While the two questions are linked, they are so complex that it is not feasible to deal with the two in a single chapter. The sharing of the benefits issue is addressed in detail in Chap. 5, in which the reader is introduced to characteristic function games. Chapter 6, turns to the stability issue, emphasizing the constant threat to RFMOs of defection and free riding. In so doing, the chapter introduces the reader in turn to partition function games.[4]

Chapter 5 examines characteristic function games in detail, leading to an analysis of fair sharing rules, with the rules discussed being: the Nash Bargaining Solution, the Shapley Value and the Nucleolus. It is concluded that side payments play a crucial role to ensure the stability of international fisheries agreements.

Chapter 6, in turn, delves into partition function games. These games model the formation of international fisheries agreements endogenously, and hence can be used to predict the exact agreement that will form. The results show that the likelihood of achieving a stable agreement among all fishing states decreases significantly with the number of players involved in the fishery. It is also demonstrated that the success of cooperation decreases with the efficiency of the fishing fleets, and increases with their asymmetry.

The policy conclusions arising from Chap. 6 are straightforward. RFMOs must do everything in their power to curb the so-called unregulated fishing in the high seas under RFMO jurisdiction. Unregulated fishing, which is uncurbed, will serve to undermine the RFMO, leading the RFMO cooperative game to degenerate into a non-cooperative one.

---

[4]In the discussion of non-cooperative games in fisheries, the point was made that cooperative games are linked to non-cooperative games. Chapter 6 provides a clear example with its title being: "Non-cooperative Coalition Formation Games".

8.2 Future Challenges

The book, thus far, has, it is hoped, demonstrated the fact that the relevance and importance of the concept of strategic behaviour to fisheries management has steadily and significantly increased over the past several decades. It is hoped, as well, that the book has made evident the great strides that have been made in developing the game theoretic analysis applied to fisheries management over this time period. While much has been accomplished over the past forty years, the authors, without hesitation, concede that many challenges, to be addressed by future researchers, still remain. In this subsection, several of these key future challenges are presented. The authors make no claim that the list is exhaustive.

8.2.1 *Uncertainty*

A key future challenge comes under the broad heading of uncertainty. In the preceding chapters, the game theoretic models presented have been largely deterministic ones. There is no questions that such models are of great importance for expositional purposes. Having said that, however, the real world of ecology and policy is one of irreducible uncertainty. Global warming is one obvious cause of uncertainty, but it is only one of many. Uncertainty carries with it the question of the resiliency of international fisheries agreements, a question raised in Chaps. 1 and 4. Outside of saying the obvious that, while uncertain events cannot by definition be predicted they can be anticipated, the present game theoretic analysis has little to offer the international fisheries agreement (IFA) policymaker.

Although uncertainty has not been used in coalition formation games that model the formation of IFAs, it has been introduced in other parts of the fisheries economics literature. Ellefsen et al. (2017) introduce ecological uncertainty into a simultaneous move game that is not a coalition formation game. Specifically, there is uncertainty about the migration pattern of the fish stock. The authors show that this ecological uncertainty may lead to the breakdown of IFAs. Another strand of the literature has also investigated how uncertainty from climatic changes may affect fishing decisions of fishermen/fishing states (e.g. Mina et al. 2016). Using field experiments among fishermen from Mexico and the Columbian Pacific, Mina et al. (2016) find that removing the uncertainty about the consequences of a climatic event causes fishermen to decrease catches. The results of their paper thus suggest that climate-related uncertainty leads to the overexploitation of internationally shared fish stocks. Generally, uncertainty is an underexplored topic in the fisheries economics literature.

8.2.2 Evolutionary Games

The second area of future challenges is that of evolutionary games. Most games are in reality repeated through time. Players, if they are rational, will build up their stores on knowledge—learn—which will have an impact on future games. Repeated games introduced into the fisheries economics literature over 20 years ago (Hannesson 1997) provide us with a start, but they are not sufficient. There is, it is true, an extensive literature on evolutionary games in the field of biology. It is also true that there have been attempts to adapt these games to economics (e.g. Friedman 1991). That being said, there have, as yet, been no attempts to apply evolutionary game theoretic analysis to fisheries economics. The authors take the view that attempts should be made without further delay.

8.2.3 Ecosystem Games

Many of the presented models consider only a single species. Only in a few models, we have multi-species fisheries, mostly presented in Chap. 7, but there is a potential in making the approach even broader to a full ecosystem game. Such analysis should consist of different fisheries with biological and/or economic interactions. The intention is to incorporate, for example, the role of salinity, ocean currents, eutrophication and the whole ecosystem in general in the model. The ecosystem games naturally lead us to the next issue of future challenges, namely that of multi-sector games.

8.2.4 Multi-sector Games

Most of the applied game theoretical models are relevant for commercial fisheries, seen in isolation. In point of fact, the fishing sector, in other than unusual circumstances, is to be seen as being linked to other sectors. For example, commercial fisheries are often closely linked to recreational fisheries. Furthermore, the fisheries sector can have a linkage with the power industry in terms of wind power or hydropower. For instance, dams affect migratory fish stocks, with salmon being a prime example (Nieminen 2017). There are many other links between the fishing sector and other sectors, such as aquaculture, processing sector, food chains, consumers and NGOs.

8.2.5 National/Regional Fisheries Management and Game Theory

We deem the application of game theory to national/regional fisheries management to be a future challenge of particularly great importance. In Chap. 1, the point is made that, while game theory has been applied to national/regional fisheries management to a limited degree,[5] the application has lagged far, far behind the application of game theory to the management of international fisheries.

The case study in Chap. 7 on the groundfish trawl fishery of British Columbia provides an indication of just how much needs to be done. In national fishery games, there is of course the issue of cooperation among fishers, similar in some ways to cooperation among fishing states in international fisheries. But, unlike international fisheries, and as emphasized by the Kronbak–Lindroos model discussed in Chap. 7, there is a key "third" party, namely, the national resource manager. As the aforementioned case study, makes clear, a national cooperative fisher game does not, as a consequence, have to be self-enforcing, in contrast to international fishery games. Moreover, and of great importance for resource management, there exists the possibility of multi-level cooperation; cooperation among fishers combined with cooperation between fishers and resource managers. In addition, a national fishery game can be multi-sector in nature; it can have ecosystem elements and can be evolutionary in nature. Finally, uncertainty plays a major role.

In other presentations, the authors have put forth the claim that the application of game theory to national/regional fisheries management is to be seen as the New Frontier (e.g. Grønbæk et al. 2019). We repeat the claim.

8.3 Concluding Remarks

In Chap. 1, the point is made that the impact of game theory upon modern fisheries economics, from modern fisheries economics inception in the mid-1950s until the late 1970s, was negligible. It is hoped that this book, through its discussion of the application of game theory to fisheries management and its policy relevance, accompanied by many case studies, has made it abundantly evident that game theory is now the tool indispensable for the analysis of the economic management of international fisheries. One simply cannot undertake serious analysis of the economics of international fisheries management, without applying game theory.

The application of game theory to the economic management of domestic fisheries lags behind its application to international fisheries. That being said, it is but a matter of time, the authors contend, before game theory is seen to be as indispensable to the analysis of domestic fisheries management, as it is to the management of international fisheries.

---

[5]See, for example, Sumaila (2013) and Bellanger et al. (2019).

The preceding chapters demonstrate that the accomplishments over the past 40 years have been impressive. Nonetheless, much remains to be done. The scope for future research in the application of game theory to the economic management of fishery resources, domestic and international, is immense.

References

Bellanger, M., Holland, D., Anderson, C., & Guyader, O. (2019). Incentive effect of joint and several liability in fishery cooperatives on regulatory compliance. *Fish and Fisheries, 20,* 715–728.

Clark, C. W. (1980). Restricted access to common-property fishery resources: A game theoretic analysis. In P. Liu (Ed.), *Dynamic optimization and mathematical economics* (pp. 117–132). New York: Plenum Press.

Ellefsen, H., Grønbæk, L., & Ravn-Jonsen, L. (2017). On international fisheries agreements, entry deterrence, and ecological uncertainty. *Journal of Environmental Management, 193,* 118–125.

Friedman, D. (1991). Evolutionary games in economics. *Econometrica, 59*(3), 637–666.

Grønbæk, L., Lindroos, M., Munro, G., & Pintassilgo, P. (2019). *Game theory and fisheries management: Insights and future areas of research.* Paper presented to the North American Association of Fisheries Economists Forum 2019, Halifax, Nova Scotia, May 2019.

Hannesson, R. (1997). Fishing as a supergame. *Journal of Environmental Economics and Management, 32,* 309–322.

Levhari, D., & Mirman, L. J. (1980). The great fish war: An example using a dynamic Cournot-Nash solution. *The Bell Journal of Economics, 11*(1), 322–334.

Mina, J. S. A., Fernández, D. A. R., Ibarra, A. A., & Georgantzis, N. (2016). Economic behavior of fishers under climate-related uncertainty: Results from field experiments in Mexico and Colombia. *Fisheries Research, 183,* 304–317.

Nieminen, E. (2017). *Bioeconomic and game theoretic applications of optimal Baltic Sea fisheries management—Towards a holistic approach.* Ph.D. thesis, University of Helsinki, Department of Economics and Management, Publications Nr. 65.

Sumaila, U. R. (2013). *Game theory and fisheries: Essays on the tragedy of free for all fishing.* Abingdon: Routledge.

Printed in the United States
By Bookmasters